DATE DUE	DATE RETURNED
DEC 1 8 1992	OCT. 0 2 REC'D
DEC 0 6 1993	NOV 2 0 REC'D
MAY 1 3 1994	OCT 2 0 REC'D
NOV 1 9 2000	OCT 3 0 REC'D
DEC 1 3 2001	DEC 2 0 REC'D
MAR 3 1 2003	APR 2 1 REC'D

D1165703

Grassland Ecology and
Wildlife Management

Grassland Ecology and Wildlife Management

E. DUFFEY
Head of the Lowland Grassland Research Section,
Monks Wood Experimental Station

M. G. MORRIS, J. SHEAIL, LENA K. WARD,
D. A. WELLS and T. C. E. WELLS
Lowland Grassland Research Section,
Monks Wood Experimental Station,
Huntingdon

With a Foreword by
Professor K. Mellanby, C.B.E.

LONDON
CHAPMAN AND HALL

First published 1974
by Chapman and Hall Ltd
11 New Fetter Lane, London EC4P 4EE
© 1974 Chapman and Hall Ltd
Printed in Great Britain by
Cox & Wyman Ltd, London, Fakenham and Reading

ISBN 0 412 12290 1

Distributed in the U.S.A.
by Halsted Press, a Division
of John Wiley & Sons Inc., New York

Library of Congress Catalog Card Number 74 – 1550

Contents

v

Foreword

By K. Mellanby

Director, Monks Wood Experimental Station

Monks Wood Experimental Station was established in 1960 by the Nature Conservancy to study practical problems of conservation. The Station was particularly intended to supply information which would enable the Conservancy to discharge the function placed upon it by its Royal Charter of 1949 'to provide scientific advice on the conservation and control of the natural flora and fauna of Great Britain' and help the Conservancy to manage the various Nature Reserves which it acquired. Dr E. Duffey, the senior author of this book, was a member of the Working Party which drew up the plans for the new Experimental Station, and was also one of the original Section Heads appointed when its work began. At first his was called the 'Conservation Research Section', and it was intended to work on a whole series of different types of habitat. However, it was soon realized that it would be unwise to try to cover too great a range of terrains, particularly as funds and staff were not as readily available as had initially been hoped, and so Dr Duffey decided to concentrate on grassland, as it is a widespread vegetation type of great ecological interest which is particularly endangered by modern developments in agriculture and forestry.

The Lowland Grassland Research Section is a multi-disciplinary group mainly consisting of zoologists and botanists but also including a historical geographer, which has been at work for over ten years. Their research has always been of a strictly applied nature, and has had as its goal the better management and conservation, in the widest sense, of grassland. However, as in other fields of work, their research has shown that it is impossible to distinguish sharply between 'pure' and 'applied' research, and a considerable amount of new fundamental ecological knowledge, giving rise to specialist scientific publications, has emerged from their efforts. At the same time they have not been diverted from their practical objectives, and

xi

already their work has been made available to wide audiences both within and outside the Nature Conservancy by four Symposia which they have organized at Monks Wood. These have dealt with the conservation of invertebrates, on grazing as a conservation tool, on the biotic effects of public pressure, and, finally, on the ecological and archaeological importance of old grassland.

This present book is intended to bring their work before an even wider public. It will, I am sure, be widely welcomed as further evidence that research can give essential and practical help to conservation, while at the same time adhering to the high standards demanded by the professional ecologist. The book should be of considerable use to the scientists and administrators under the new Nature Conservancy Council, set up in 1973 when the old Nature Conservancy was divided into it and the Institute of Terrestrial Ecology with which Monks Wood remains. However, it should be equally valuable to many others. Today we have a growing number of people, particularly in voluntary organizations such as County Naturalists' Trusts, as well as those in national organizations and in the educational and administrative bodies, and a considerable number of private landlords, who wish to manage their property with at least some regard to the flora and fauna. And finally it should appeal to academic and practical ecologists in all parts of the world, for though the book is based mainly on British experience, it illuminates many important world-wide problems.

Kenneth Mellanby

Preface

Grasslands are one of the most widespread vegetation types in the low-rainfall areas of the world and in temperate regions where man has cleared the forest to create additional pasture. The importance of grasslands as an exploitable crop, for many amenity purposes, and as an integral part of attractive landscapes has been well studied by agriculturists, planners and landscape architects. In general, however, these interests have been concerned mainly with artificial swards created for specific purposes, and the potential value of surviving semi-natural grassland was unrecognized except by ecologists and conservationists. In addition these grasslands have been rapidly reduced in area by reclamation for other use, by succession to scrub in the absence of grazing, or by conversion to a different type through agricultural improvement. This loss has been particularly great on those chalk and limestone soils which are easily worked and in the valley meadows of lowland rivers where the use of fertilizers and herbicides boosts productivity but at the expense of floristic and faunistic diversity.

Although permanent pasture and rough grazing (mostly unenclosed in the upland regions of the north and west) cover well over half the land surface of England, Wales and Scotland only a very small proportion is at present included in the series of National Nature Reserves. For example, there are only eight chalk grassland National Nature Reserves with a total area of 1136 ha. Representation of other types of calcicolous grassland is even poorer; three National Nature Reserves totalling 142·0 ha have been established on Carboniferous limestone, one on oolitic limestone (25·0 ha) and none on Devonian or magnesian limestone. Meadow grasslands occur on only two National Nature Reserves, Upper Teesdale and at Cricklade (Wiltshire). However, proposals have been made which should improve this situation. Elsewhere in southern England permanent

xiii

grassland of biological interest survives on non-agricultural land such as public open spaces, some National Trust properties, and County Naturalists' Trust nature reserves, roadside verges and commons as well as on the grazing lands of the downs and on lowland pastures and hay meadows.

In 1968 the Lowland Grassland Research Section at the Monks Wood Experimental Station had an opportunity, during the Nature Conservancy's comprehensive ecological survey for the Nature Conservation Review, to take stock of the surviving areas of lowland grassland which retain some scientific interest. The Survey was the most ambitious attempt by any country to assess its wildlife resources, to bring together available ecological knowledge and, on the basis of this information, to decide on priorities for protecting the best sites.

Five years earlier the Section had begun a series of long-term studies on the plants and animals of the grassland ecosystem with special reference to the responses of certain species to experimental grazing and mowing. Later, the investigations were expanded to include ecological studies on scrub plants which colonize grasslands during succession together with their associated fauna. In addition to the biological work, complementary historical studies were made of the land-use factors which have modified grassland plant and animal communities, particularly during the last two to three hundred years. This book, therefore, is mainly an attempt to present, in a single integrated account, the historical, botanical and zoological studies made by the Lowland Grassland Research Section on the development, range of variation, scientific interest and management of semi-natural grasslands in the lowland areas of England. The results are set in the context of available knowledge on grassland ecology drawn from many sources in this country and from other parts of the world. Our conclusions will, we hope, have both theoretical value for other ecologists and practical use for conservationists and land managers responsible for grassland maintenance.

From the commencement of our joint studies we recognized that in addition to their floristic and faunistic diversity conditioned by soil type, water regime, geographical location and other factors, semi-natural lowland grasslands have been greatly modified by human influence. Through centuries of exploitation the vegetation and animal life we see today have evolved as much in relation to man's activities (although very imperfectly understood) as through the physical environment. In those few areas where grasslands can be regarded as natural, or nearly so, modification has been slight or negligible so that wildlife management should consist mainly of preventing disturbance. On the other hand in semi-natural areas, where the potential natural vegetation would be very different in the absence of man's influence, some controlled intervention must often be adopted if the special conservation interest of the area is to be maintained. Several alternatives are usually available. If detailed information about the site is rather sparse then the best policy may be to continue the method of traditional land use which maintained the scientific interest in the past. In some cases a modification of this practice may be necessary in order to improve the conditions for particular species of plants and animals. In other areas there may be a strong scientific reason for allowing natural succession to take place so that the rate and nature of change can be studied. Such courses of action depend largely on the objects of management for individual sites and it is important that these should be defined before management policies are formulated.

Many countryside areas, whether by custom or design, serve a dual purpose, and this has led to the development of the planning concept 'multiple use'. It has been widely applied to nature reserves, although sometimes in a misleading way. For example, although many reserves, particularly those on coastal sand dunes, dry heathlands and chalk downs, are popular with the public for outdoor activities, the importance of their main function – to conserve wildlife – is not lessened even though effective management may be made more difficult. Growing public interest in the natural environment and an increasing population with greater mobility clearly indicate that in spite of the countryside amenities now available the demand for access to nature reserves will increase as open, pleasant landscapes become a diminishing resource. The tolerable level of public use of an area varies according to the terrain and vegetation cover (Plate 1); for example, most grasslands are more resistant than a lichen-covered dune which soon breaks up, or wet ground which is reduced to a muddy patch after only light trampling.

Related to this aspect of the use of nature reserves is the question whether economic exploitation is compatible with wildlife conservation. For instance, agricultural or forest products sometimes become available on nature reserves as a result of scientific management. Examples are the use of sheep and cattle to achieve the objectives of conservation (discussed in detail in this book), the production, from time to time, of coppice and thinnings on some woodland reserves, and the cropping of reed and sedge on fenland reserves. Commercial gain cannot, however, be an essential part of the scientific management policy because it would soon conflict with the requirements of conservation. In later chapters it will be shown that sheep and cattle on grassland nature reserves are not usually grazed at the same stocking rate nor, necessarily, at the same seasons of the year as required by the farmer, whose aim it is to obtain maximum productivity. Neither can fertilizers and pesticides be used, as in good farming practice, because they would greatly modify the plant and animal communities and perhaps cause extinction of some species.

Reference has already been made to some ways in which conservation has social value in modern life, but it is sometimes asked whether the preservation and study of wildlife are scientifically important in the sense that they make a contribution to economic problems in agriculture and forestry and help to maintain a healthy and pleasant environment in which to live. In other words, is wildlife conservation a pleasure or a necessity, or both? Most conservationists believe that unspoilt landscapes and the wildlife associated with them are part of our national heritage, of equal importance with the greatest achievements in the arts, because of the pleasure they give to so many people. This belief, they would say, is sufficient to justify government action to protect and preserve them for the nation. Nevertheless research on conservation ecology has a far wider application in scientific affairs than learning how to maintain populations of attractive and interesting plants and animals. Our knowledge of the biology, distribution and ecology of plant and animal species is almost insignificant in relation to the number of species described. Over 900 000 species of insects are known and about 215 000 other animals, and yet only a few species, mostly pests and others of commercial importance, have been studied in any detail. Similarly the world total of higher plants is between 250 000 and 300 000 species, of which well under 1% are known at present to be of commercial value, and yet the biology of even this small proportion is still very inadequately studied.

To allow extinction of plants and animals to take place in our present state of knowledge, through short-sighted planning and land management, is a poor investment of scientific expertise. It must also be remembered that the science of ecology is still in an early stage of development and it is only in recent times that its importance has been recognized by the agriculturist, forester, landscape architect, economist and planner. For example the widespread problem of restoring industrially degraded environments, both terrestrial and aquatic, concerns all of these disciplines. Each relies, in different ways, on ecological expertise although the facts available lag behind the demand for urgent practical action. In this case one can look forward to environmental improvement in the future and an increase in relevant ecological research but other trends in the countryside of most 'developed' countries are causing concern. Modern agricultural techniques, closely linked with economics, lead to monocultures of crops and increasing field size, while habitat diversity in the landscape is reduced by hedge clearance, ploughing verges, filling in ponds and the widespread use of pesticides. Conservation can help to reduce the impact of this trend by creating nature reserves, encouraging the preservation of land-scape features such as trees, hedgerows and clean waterways and so contributes to the maintenance of a more varied and attractive environment with a richer wildlife.

Scientific conservation is therefore applicable to a wide range of prob-lems, both natural and man-induced, but in every case the same basic data are required; an understanding of the interactions between plants and animals in wildlife communities and their responses to measurable factors in the physical and biotic environment.

The conservation of nature is, in effect, another way of describing the conservation of man: ultimately his survival will depend on a full under-standing of nature and an ability to apply this knowledge to his own well-being.

Monks Wood Experimental Station,   E.D.
(Institute of Terrestrial Ecology),
Huntingdon
April 1974

Authors

Although the six authors have contributed to all parts of the book by discussion and comment, the chapters reflect the special knowledge of individual members. The principal authorship of the chapters is as follows:

Chapter 1. J. Sheail, T. C. E. Wells, D. A. Wells and M. G. Morris
Chapter 2. T. C. E. Wells and D. A. Wells
Chapter 3. M. G. Morris
Chapter 4. D. A. Wells
Chapter 5. M. G. Morris and E. Duffey
Chapter 6. L. K. Ward
Chapter 7. T. C. E. Wells
Chapter 8. T. C. E. Wells
Chapter 9. T. C. E. Wells
Chapter 10. T. C. E. Wells
Chapter 11. L. K. Ward
Chapter 12. M. G. Morris

Acknowledgements

During the course of our work we have received much help and advice from our colleagues in the Nature Conservancy, both at Monks Wood and elsewhere, particularly N. E. King, H. J. Williams, Dr G. Peterken, K. Lakhani, J. M. Schofield, Dr M. D. Hooper, J. Heath, M. Skelton, Dr J. M. Way and Dr B. Green. P. Wakely took many of the photographs and gave us every assistance. We are indebted to Professor K. Mellanby for his interest and help throughout the preparation of the book.

Other members of the Lowland Grassland Research Section, namely our assistants Heather Brundle, Mary Grimes, Karen Jefferies, Mary Maley and Peter Tinning, and past assistants Sue Freestone, Ian Gauld, Janet Essex, Sally Murrell, Geoffrey Tew and Jean Williamson, have given us valuable help with fieldwork and in other ways. We would also like to acknowledge a special debt to our secretary Gillian Searle (Mrs G. Tew) for typing and retyping the manuscript and for countless tasks in connection with the preparation of the book, to Shirley Jackson who drew all the diagrams and maps, and to Mrs Rita Duffey who prepared the index and helped with proof-reading.

For help and advice, or obtaining material for the text, we are indebted to Dr J. Berry, C. A. Elbourn, F. H. W. Green, T. Larsson, Dr W. J. Le Quesne, Dr P. Lloyd, G. H. Mayhead, E. C. Pelham-Clinton, the late F. Reynolds, Dr. I. Rorison, the Society for the Promotion of Nature Reserves (Association of Nature Conservation Trusts) and in particular A. E. Smith, Dr M. G. Solomon, D. T. Streeter, Dr N. Waloff and G. E. Woodroffe.

We are indebted to the following for permission to reproduce plates and figures: Animal Ecology Research Group (University of Oxford); S. Beaufoy; Dr J. Berry; Miss Jane Burton; Cambridge University

Collection; Dr B. Forman; H. M. Frawley; Geographical Survey Office of Sweden; J. A. Grant; Dr J. L. Mason; Ministry of Defence (Crown Copyright); Dr M. C. F. Proctor; J. Robinson; P. Stuttard; P. Wakely; T. C. E. Wells; Academic Press, London, Table 5.1; J. R. Aldous and Blackwell Scientific Publications, Oxford, Table 11.1; Dr G. W. Arnold and Blackwell Scientific Publications, Oxford, Fig. 8.1; Blackwell Scientific Publications, Oxford, Figs. 4.2, 7.5 and 8.4; Grassland Research Institute, Hurley, Fig. 7.2; Dr J. L. Harper and Centre for Agricultural Publishing and Documentation, Wageningen, Fig. 7.1; Professor D. H. Janzen and the American Naturalist, Chicago, Fig. 6.3; Dr R. M. Morris and the Grassland Research Institute, Hurley, Figs. 7.3 and 7.4; Dr M. J. T. Norman, Fig. 9.1; W. B. Saunders Co., Philadelphia, Figs. 3.3 and 3.4; Dr D. W. Shimwell, Fig. 2.2; K. C. Side and the Amateur Entomologists Society, Fig. 6.6; Dr C. J. Smith, and Blackwell Scientific Publications, Oxford, Figs 10.1 and 10.2; Professor R. Whittaker, Fig. 3.1; Dr T. J. Williams, Fig. 4.1; Dr M. Williams and the Institute of British Geographers, Fig. 1.6; K. Williamson, R. Williamson, and H. F. & G. Witherby Ltd, London, Fig. 6.8.

Nomenclature

The nomenclature adopted is derived from the following:

Birds: Hudson, R. 1971. *A Species List of British and Irish Birds.* British Trust for Ornithology. Guide no. 13.

Mammals: Southern, H. N. (Ed.) 1964. *The Handbook of British Mammals.* Blackwell, Oxford.

Spiders: Locket, G. H. & Millidge, A. F. 1951–3. *British Spiders, Vols I and II.* Ray Society, London.

Insects: Benson, R. B. 1951, 1952, 1958. *Symphyta. Handbk Ident. Br. Insects,* VI (2).

Kloet, G. S. & Hincks, W. D. 1945. *A Check List of British Insects.*

Kloet, G. S. & Hincks, W. D. 1964. *A Check List of British Insects. Second Edition (Revised) Part I: Small Orders and Hemiptera.* Royal Entomological Society, London.

Kloet, G. S. & Hincks, W. D. 1972. *A Check List of British Insects. Second Edition (Revised) Part 2: Lepidoptera.* Royal Entomological Society, London.

Waloff, N. 1968. Studies on the insect fauna on Scotch broom *Sarothamnus scoparius* (L.) Wimmer. *Adv. ecol. Res.,* **5**, 87–208.

Flowering plants: Clapham, A. R., Tutin, T. G. & Warburg, E. F. 1962. *Flora of the British Isles. (2nd Edn.),* Cambridge University Press.

I Grasslands and their history

The vegetation cover of Britain under natural conditions would be forest, yet today grasslands account for about 61% of the total agricultural area. How has this change come about? This chapter will examine the distribution and management of grasslands in the past, particularly the many factors which have combined to create them, and their use and modification by the farmer.

Sources of Information

Man has always been interested in grasslands, their extent and management, because they are a vital source of food for his livestock. As a result, there is a great deal of printed evidence on the meadows, pastures and downlands of the past. Old farming books are of special importance, particularly the volumes of the Board of Agriculture, published between about 1790 and 1820, which have been intensively used by agricultural historians. The Board, while preparing the reports, distributed questionnaires to many 'intelligent farmers', and incorporated the replies in the reports which were later published for each county. These volumes, together with other printed works, give a useful survey of the character and distribution of grassland at that time. But it should be remembered that writers tended to concentrate on the larger, more progressive farms, which may not have included sites of particular interest to botanists and zoologists today. Furthermore, very few printed books identify the use and management of individual fields and open spaces. The Board of Agriculture report for Oxfordshire (Young, 1813), for example, contains an excellent survey of land use in the Thames valley, but it is impossible to deduce the use and management of such sites as the

1

meadows of Pixey and Yarnton Meads, which are of great interest to the botanist.

There are, of course, many documents in the county record offices and private archives which give a more detailed account of the grasslands in individual parishes and districts. Where these manuscripts survive and can be found, they are of great value in reconstructing the history of sites and the probable vegetation cover, but it is worth noting that many farmers, shepherds and graziers, with an intimate knowledge of the grasslands of the past, could not read or write, and so they were extremely unlikely to leave any record of their observations! Even those who *were* literate have left few records of interest to the botanist and zoologist. Very few books and papers noted the presence of such plants as the pasque flower or the monkey orchid because this kind of information was hardly relevant to the economic well-being of a farm or estate. This is in contrast to woodlands where the names of trees were often recorded because the species varied in their economic value. The historical study of grasslands, therefore, presents rather more difficult problems than some other types of vegetation.

There was no systematic attempt to collect statistics related to land use until the mid-nineteenth century, and the first of the now familiar Annual Returns was made in 1866, providing information on crops and livestock (Ministry of Agriculture, 1968). Since 1892, holdings of at least 1 a. (0·4 ha) have been included in the annual survey. A distinction has usually been made between *permanent grass, temporary* or *rotation grass,* and *rough grazings.* At this point it is helpful to define each of these terms as frequent reference will be made to them in later chapters.

Permanent grass is usually over 7 years old and is normally enclosed and used for pasture or hay. It is relatively uniform in structure and does not contain tussocky species, nor shrubs, such as heather (*Calluna*). It is found mostly in the lowlands of England and Wales, although some permanent grass can be found along rivers and streams in upland regions.

Temporary grass is made up of a grass-clover mixture which is sown as part of the arable rotation. It is ploughed and returned to arable within a period of 3 to 5 years. This type of grassland is often described as *ley* in the agricultural statistics.

Rough grazings include land of a most varied character, and it is not always easy to decide whether a particular grassland should be allocated to the permanent grass or rough grazings category. Generally speaking, they are made up of native species which have not been planted by man, often including those areas which are described by agriculturists as marginal, unimproved or uncultivated. Rough hill pastures of moorland and upland Britain, lowland heaths and moors, and many unenclosed commons and rough marsh pastures fall into this category. In addition, arable fields which have reverted to grassland during a period of agricultural depression are often described as rough grazings.

The Annual Returns of the Ministry of Agriculture are of considerable value in assessing the changes in the area of grasslands since 1866, but they are of no use in tracing changes in the proportion of land under grass on individual farms. The Ministry allows the parish returns to be inspected at the Public Record Office, but those related to farms and holdings remain strictly confidential. The Returns, therefore, are of no help in tracing the history of individual fields and open spaces, and this has severely impeded the work of the historian.

Old maps are an important source of information where they indicate the presence, extent and layout of grasslands. But there are less than 30 extant maps for the period before 1500 (Harvey and Skelton, 1969) and many early maps were published at such a small scale that they are of little use to the biologist undertaking large-scale studies of particular sites. For example, the maps prepared by John Rocque of London and by Chapman and André of Essex give a general impression of land use in those parts during the eighteenth century, but they cannot be used to discover the use of individual places. The Ordnance Survey was established in 1791, but few large-scale maps were published before 1840 (Harley, 1964). For a few years, the Ordnance Survey published booklets with a brief description of the use of each parcel of land marked with a number on the 1:2500 maps, but this practice stopped in 1884 and since that time the Survey has not recorded land use in any detail.

Estate maps are particularly useful. Many large landowners commissioned surveys of their property as part of a modernization scheme: the maps often helped in deciding the layout of new fields or farms, and how land husbandry should be improved. A careful watch must be kept for errors on the maps, both accidental and intentional. Comber (1772), for example, complained of landlords who bribed their surveyors into exaggerating the area of some fields and farms so that the tenants could be charged higher rents!

Maps were also required when common lands were being enclosed and redistributed into ring-fence holdings. Fig. 1.1 is based on a survey commissioned for this purpose by the landowners of Donnington, Buckinghamshire, in 1803.[1] It is particularly interesting because the surveyor described one plot as Frogcup Meadow. The word Frogcup is a local

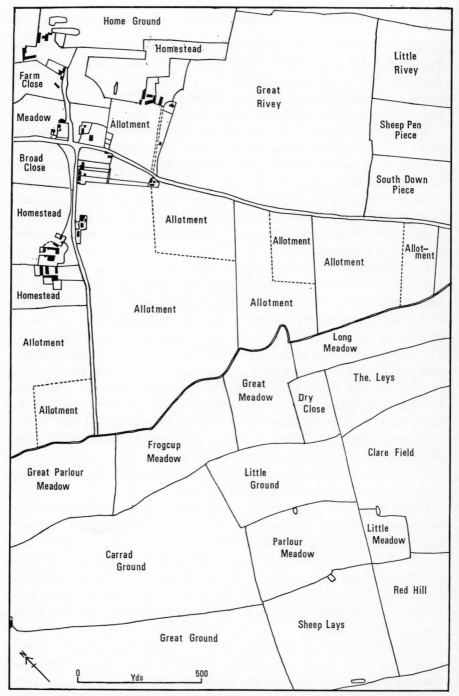

Fig. 1.1 Frogcup Meadow and the field names of part of Donnington parish, Buckinghamshire, as indicated by an enclosure map of 1803.

variation of fritillary, and this is the first reference to the plant on the site. The fritillary grew in the meadow until the early 1950s when the ground was ploughed up and the plant became locally extinct. Such references help to supplement and locate the data gleaned from early floras and other sources: the maps play an essential part in reconstructing the one-time distribution of grassland communities.

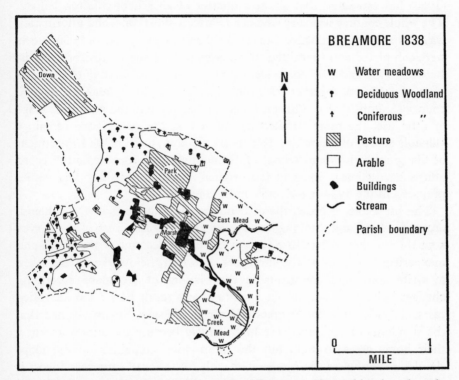

Fig. 1.2 Pattern of land use in the parish of Breamore, Hampshire, based on the Tithe Commutation survey of 1838.

Historians have also made considerable use of maps prepared for the Tithe Commutation Survey of the period 1836 to about 1850. As a preliminary to commuting all tithes to money payments, a large-scale survey was made of those areas liable to tithes (Prince, 1959), and Fig. 1.2 is based on the map and apportionment of the 'strip parish' of Breamore in Hampshire.[2] The map indicates the extent of the water-meadows on the river Avon and their spatial relationship to the pastures near the village and on the high downland. The meadows were a distinctive grassland community (see page 34), preserved for their hay and for grazing the sheep and cattle which spent the remainder of the year on the pasture and arable elsewhere in the parish.

Between 1930 and 1938, a Land Use Survey was made of every part of England, Scotland and Wales. It illustrates the value and limitations of surveys not specifically intended for the botanist and zoologist working at a large scale. The work was carried out by volunteers, and Stamp (1960) has estimated that about a quarter of all school-children helped. The whole country was surveyed in a relatively short time, and information was recorded on Ordnance Survey 1:10 560 maps. About 15 000 sheets were completed and submitted to the organizers of the survey for analysis. Between 1930 and 1945 maps showing the use of land were published at a scale of 1:63 360, together with monographs describing land use in each county (Stamp, 1962). Whilst ecologists generally find the published maps of little use, the more detailed field sheets at the larger scale remain a valuable source of reference. The sheets are now stored in the Department of Geography, London School of Economics, and some county record offices have copies. Some of the original sheets are missing, but copies prepared in the 1930s survive for most counties.

The surveyors recorded the distribution of arable, meadow, heathland, orchards, woodland, buildings and gardens. The organizers of the survey kept the number of land-use categories low on account of the youth and inexperience of many of the surveyors. However, fieldworkers were asked to write notes and comments in the margins of the maps describing changes in land use and management and as a result they often identified cases of fields reverting to grass and scrub invasion. Stamp claimed that the standards of accuracy were high, largely because the surveyors knew their home areas so well, but the information contained on the maps should, nevertheless, be interpreted with care. Some surveyors found it difficult to distinguish long fallow from more permanent grassland, and others encountered problems in defining the boundary between scrub and woodland. The organizers of the survey were aware of these problems, and checks were made and field sheets amended.

The aim of the Grassland Survey of 1939–40 was to characterize permanent pasture by using botanical criteria, and to use the results in identifying those grasslands which could be most easily improved for agriculture (Davies, 1941; Ordnance Survey 1952). The classification devised by Stapledon and Davies, and adopted in subsequent surveys of 1947 and 1959, is as follows:

First grade rye-grass pasture in which *Lolium perenne* contributed 30% or more to the sward.

Second grade rye-grass pastures containing 15–30% *L. perenne*.

Agrostis with rye-grass pastures, basically *Agrostis*, but *L. perenne* contributing up to 15%.

Ordinary *Agrostis* pastures containing traces of rye-grass.

Agrostis pasture with excess of rushes and sedges.

Festuca pasture, including mountain fescue, open downland, and lowland heaths with fescue swards.

There are in addition five grades of hill pasture which are recognizable by their dominant grass or sedge.

The survey of 1939–40 can be criticized on the grounds that it sampled only a small proportion of the permanent pasture, and that some rather broad generalizations were made. Nevertheless, it provided valuable information on the agricultural quality of British grasslands. Two maps (1:625 000) covering England and Wales show the distribution of the different categories. First and second class rye-grass pastures made up only a small proportion of the total area at the time of the survey, and they were almost entirely absent from such counties as Bedfordshire. Even where they were present, they constituted only a small fraction of the total grassland area. The data published for Huntingdonshire (Fryer, 1941) are representative of many other Midland counties with extensive areas of clay soils, demonstrating the importance of the *Agrostis* and *Festuca* pastures (see Table 1.1).

Table 1.1 The results of the Grassland Survey of Huntingdonshire. 1939–40 (*after* Fryer, 1941)

	Area occupied		Percentage of grassland area occupied
	hectares (ha)	acres (a.)	
a First class rye-grass pastures	500	1 300	1·4
b Second class rye-grass pastures	1 500	3 800	4·2
c *Agrostis* with rye-grass pastures	7 200	17 900	19·8
d & f *Agrostis* pasture with fescue pasture	25 400	62 800	69·7
e *Agrostis* pasture with excess of rushes and sedges	1 800	4 400	4·9
Total	36 400	90 200	100·0

At the time of the last survey in 1959, first and second class rye-grass pastures occupied only 19·5% of all permanent grassland excluding rough grazings. *Agrostis*/rye-grass, *Agrostis* and *Festuca* pasture, accounted for 66·3% of the area. The remaining 14·2% was made up of *Molinia*/*Nardus* moorland, and pasture dominated by rushes.

Distribution of Grasslands

The importance of permanent grasslands and rough grazings is clearly shown in the Agricultural Returns of 1971 (Ministry of Agriculture, 1972; Department of Agriculture, 1972). Of a total area of 43·8 million a. (17·7 million ha) in agricultural use, 25% are described as permanent grassland, 36% as rough grazings, and 39% as arable. There are, however, striking variations in the distribution of these three categories of land use. Only 1·0 million a. (0·4 million ha) of permanent grassland occur in Scotland, which has over two-thirds of the rough grazings, whereas there are 7·7 million a. (3·1 million ha) of permanent grassland in England and Wales, which have only 4·7 million a. (1·9 million ha) of rough grazings, largely confined to the mountainous parts of Wales.

Fig. 1.3 shows the proportion of agricultural land occupied by permanent grassland and rough grazings in each county. Eastern England has the smallest percentage of land in these two categories: this is the driest area, with few hills, and no land over 1000 ft (305 m). Permanent pasture and rough grazings occupy less than 20% of the farmland in 10 of the counties. In a few parts, notably the Holland Division of Lincolnshire, the proportion is as low as 5%, which is largely confined to the salt marshes along the coast.

Permanent pasture and rough grazings are much more evident in the western counties and Wales, where the emphasis is on pastoral farming. Arable crops are generally found only on the better, low-lying soils: rough grazings occupy the greater part of the remaining land. Scotland may be divided into three parts: the proportion of farmland under grass is lowest in the eastern counties, and especially on the coast, the south-western counties have a higher proportion under grass, and over 80% of the land in the north-west is grassland, mostly rough grazings.

The statistics of the Ministry of Agriculture and Department in Scotland give only a general picture, and frequently the biologist wants to correlate the distribution of grasslands with various soil formations. Thus, in 1966, a special survey was undertaken of grasslands on chalk soils, and a summary of the results is given in Table 1.2. Only 3·3% of the chalkland

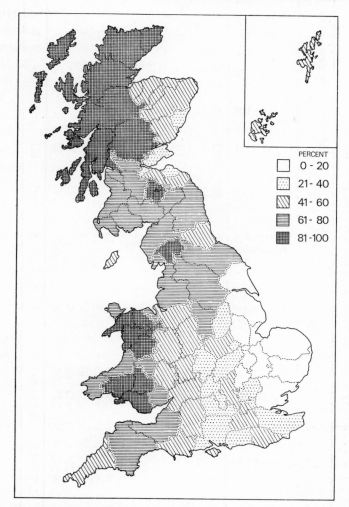

Fig. 1.3 The proportion of agricultural land occupied by permanent grass and rough grazings in 1971.

was covered by permanent pasture, of which nearly three-quarters occurred in Wiltshire, which has 14% of the total area of chalkland in England (Blackwood and Tubbs, 1970).

It is important to note that not all grasslands are situated on farmland: a large proportion of the Defence Lands is covered by rough grazings, and grass is often preserved for amenity purposes. Such landscape gardeners as William Kent and Lancelot (Capability) Brown improved and created extensive parks, with clumps and belts of trees set in a landscape of grass.

Table 1.2 Chalk Grassland Survey 1966: Summary of data obtained (*from* Blackwood and Tubbs, 1970)

County or topographical unit	Acreage* of chalk grassland	Acreage* of land on chalk	Chalk grassland percentage	Number of fragments	No. of fragments by size categories (acres*)					
					5-50	51-100	101-200	201-300	301-400	400+
Yorkshire Wolds	3 939	262 440	1·5	61	35	15	5	4	2	–
Lincolnshire Wolds	225	143 280	0·2	12	12	–	–	–	–	–
Norfolk	44	332 640	0·01	3	3	–	–	–	–	–
Cambridgeshire	170	156 600	0·1	3	2	1	–	–	–	–
Chiltern Hills (Beds., Berks., Bucks., Herts., and Oxon.)	2 028	550 080	0·4	59	47	7	4	1	–	–
Berkshire	1 573	185 400	0·8	59	50	7	2	–	–	–
Wiltshire	73 085	464 400	15·7	529	407	51	34	16	5	16
Dorset	8 371	236 880	3·5	145	101	28	10	2	1	3
Hampshire	5 224	239 760	2·2	119	95	13	7	1	–	3
Isle of Wight	2 128	14 040	15·2	24	13	6	2	–	2	1
North Downs (Kent and Surrey)	2 226	437 760	0·5	94	83	10	1	–	–	–
South Downs (Sussex)	8 592	223 200	3·8	117	83	13	11	2	1	7
Total	107 605	3 246 480	3·3	1225	931	151	76	26	11	30

* 1 acre = 0·405 ha

Even today, the grounds of Blenheim Park, Oxfordshire, contain over 2025 a. (820 ha) of grassland, trees and water, which are important for their birdlife and insects characteristic of Jurassic limestone grasslands. On a more modest scale, many estate owners converted the ploughed fields visible from their houses into grasslands. Plate 2 shows part of an eighteenth-century park at Upwood, Huntingdonshire, where the turf lies on top of ridge and furrow, shaped in the form of a reversed 'S', which indicates that the site has been under the plough during at least the medieval period (Eyre, 1955). Many of these parklands have remained under grass since their inception, and their turf is often older than any other in the locality. The vegetation has usually been intensively grazed because rough grass and scrub would have been unsightly, and the grounds were conveniently near farm buildings and milking parlours.

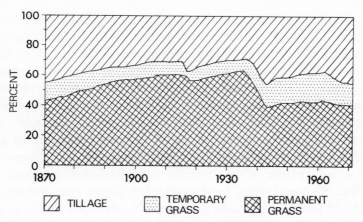

Fig. 1.4 Changes in the proportion of agricultural land under tillage, permanent and temporary grassland in England and Wales, 1871–1971.

Besides these marked regional differences, the proportion of land under grass has changed from decade to decade, and from one century to another. Fig. 1.4 is based on the Annual Returns and indicates the proportion of land under permanent grass during the last 100 years. There are, of course, many difficulties in comparing the returns of one decade with another (Coppock, 1962), but it is nevertheless clear that the proportion of land under permanent grass has varied by about 20%. The proportion was highest in the years around 1930, and there has been a relatively marked fall since about 1938. But these changes are only the most recent of a complex history in land use. Most grassland sites, especially in lowland England, were once wooded.

Prehistoric times

The origin of the vegetation of the British Isles and western Europe is outside the scope of this chapter, and the reader should consult Godwin (1956), Dimbleby (1967), Pennington (1969) and Walker and West (1970) for detailed information, while Ellenberg (1963) is a useful starting point for searching through the continental literature. These sources will also provide information on the rather specialized field and laboratory techniques required for analysing early plant remains. The history of the Quaternary fauna has also received a good deal of attention, and detailed analyses have been carried out on collections of mammal bones, fish scales, mollusca (e.g. Sparks, 1964; Evans, 1972), remains of the crustacea, and the skeletons of beetles.

The discovery and identification of plant and animal remains not only extends our knowledge of individual species, but the species themselves may indicate the character of the environment in the past. This may be illustrated with respect to Coleoptera. Older entomologists tended to assume that Quaternary beetles were different from contemporary species and, by a process of circular argument, erected numerous new taxa for extinct species but, as Coope (1968) has illustrated, they were in fact identical with modern forms. All the Pleistocene Coleoptera so far examined have been morphologically indistinguishable from recent ones. There is no evidence of any morphological evolution, let alone actual speciation. Shotton (1965), for example, has shown how even the small beetle fauna found in the Castle Eden deposits of about 1 million years ago (Tiglian age) were probably composed of species found in the environment today. The use of remains of Coleoptera as an indicator of past environments is, of course, highly complex, and it has been suggested that physiological and behavioural evolution may have occurred, unassociated with morphological change, but so far no evidence has been discovered to support that contention (Coope, 1969, 1970). With caution, Coleoptera, Mollusca and other animal remains may be used to indicate changes in the environment, precipitated by climatic and other factors.

The Pleistocene was a period of great climatic fluctuation. Conditions for long periods of the different glaciations were arctic but warmer periods occurred during the interglacials. The coleopterous fauna throughout this whole period was morphologically unchanging and responded to the climatic changes by enormous movements of populations. A movement of a few metres a year by a population of beetles advancing behind or retreating before 'the ice' was quite sufficient to account for displacements of hundreds or thousands of kilometres in the course of time. The picture painted by recent studies is of species of Coleoptera shifting bodily with their habitats in response to climatic change. About $12\frac{1}{2}\%$ of the 800 or so species recorded in Quaternary deposits in Britain no longer inhabit the British Isles (Shotton, 1970). Some can be found today only in places far removed from Britain. For instance, *Helophorus obscurellus*, for some time supposed to be an extinct species *H. wandereri*, is restricted to Siberia (Coope, 1970; Shotton, 1970). The oldest substantial faunal assemblage of Coleoptera recorded in Britain occurred at Nechells, Birmingham (Shotton and Osborne, 1965), and dated from the Hoxnian interglacial of about 350 000 years ago. A very mixed assemblage, it contained such species as the weevils *Rhyncolus lignarius* and *Rhynchaenus quercus* which are associated with trees, and others such as *Apion apricans* which are characteristic of, but not restricted to, grasslands. Many Quaternary faunas, particularly those dating from the Weichselian, do not show this diverse character, but are 'open ground' associations, developed in response to a cold climate, often with tundra-like conditions. It should be noted when analysing the data that fossil faunas tend to be more readily preserved in dried-out ponds and similar places, with the result that the remains of marshland, swamp and aquatic species are more likely to survive than those of forest and other biotopes.

The origin and age of grasslands in Europe are fertile fields for research. Although the main factors which contributed to their formation are known, there have been comparatively few detailed studies of areas under grassland today. Following the late-glacial period 14 000 B.C. to 10 000 B.C., the climate of Britain became warmer and most of lowland Britain was covered with forest during the Atlantic Period of 7000 to 5000 B.C. But from that time onwards, man began to clear substantial areas of forest, and provided opportunities for grass and herb species to spread. Godwin (1944) was the first person to draw attention to this process of prehistoric forest clearance. On the basis of pollen analyses, he found evidence of a decline in tree species and a rise in grasses and plantains at Hockham Mere, Norfolk, at about 3000 B.C. The clearings in the wood provided man with the opportunity to practise shifting cultivation and to increase food supplies for his livestock. Similar studies of pollen diagrams are being undertaken elsewhere in Britain, often supported by radio-carbon dating of critical horizons, and gradually the broad pattern of the vegetation during that period is being reconstructed. The faunal evidence matches very well this pattern of forest clearance. As far as can be judged grasslands are very 'new' biotopes, geologically speaking. Unwittingly, man created, by his management of agricultural grazing animals, communities of plants and animals unlike any previously known.

The process of forest clearance gathered momentum during the Neolithic, and by 1700 B.C. man had greatly extended many of the woodland glades previously maintained by wild herbivores (Iversen, 1949). According to Troels-Smith (1960), man supplemented the fodder of his livestock with the shoots and leaves of selected trees and shrubs, showing a preference for the elm. Archaeological evidence suggests that Neolithic man made his greatest impact on chalkland soils, so that by the Bronze Age of 1700 to 400 B.C. much of the chalkland of Britain and France was under cultivation. The many thousands of burial mounds, relict field systems and other monuments suggest that most of the woodland on the chalk had been felled. This is confirmed by the analysis of pollen and mollusca remains found at Brook, Kent (Kerney, Brown and Chandler, 1964). There may have been two phases in woodland clearance in the vicinity of Brook, the first during a transitional phase between the late Neolithic and early Bronze Age, the Beaker period, and the second during the Iron Age 'A' period. The reasons for these changes may be extremely complex, for the clearing of the wood for arable farming may have coincided with the sub-Atlantic deterioration of the climate.

Regional differences in the incidence and rate of forest clearance may be illustrated on the basis of evidence from two deposits containing a rich collection of beetle remains. A deposit of about 3300 years old was found in an artificial shaft at Wilsford, Wiltshire (Osborne, 1969), which contained a large number of remains of dung beetles (e.g. *Onthophagus*, *Geotrupes* and *Aphodius* spp.) and phytophagous Coleoptera, such as root-feeding Scarabaeidae and various species of Chrysomelidae and Curculionidae, characteristic of grassland or open ground and herb communities. The fauna clearly indicates the presence of a grassland, or grassland-like plant community, with large numbers of grazing animals. This association of beetle species is very different from that described from deposits of roughly similar age (about 1100 B.C.) at Thorne Moor, Yorkshire (Buckland and Kenward, 1973). At this site, an entirely forest fauna was recorded, containing several species which are either very restricted or do not exist in Britain today, such as *Prostomis mandibularis* (Cucjidae), *Zimioma grossa* (Ostomidae) and *Dryophthous corticalis* (Curculionidae).

Palynological evidence also suggests that progress was much slower outside the chalkland areas. The low frequency of herb and grass species suggests that most of the clearings were only temporary, and that regeneration soon took place (Turner, 1965). But from the Iron Age of about 400 B.C. onwards, and following the development of more efficient and easily produced tools, the clearance of the woodland become more widespread and permanent. Disforestation had occurred in parts of Wales by 400 B.C. (Godwin and Willis, 1962) and parts of the Humber region by 300 B.C. (Turner, 1962). A similar trend has been detected on the basis of palynological and archaeological evidence on Dartmoor (Simmons, 1964), around Malham Tarn (Pigott and Pigott, 1963), and in south Norfolk (Godwin, 1967).

Kelly and Osborne (1964) have described the chronology of development at Shustoke, Warwickshire, where deposits of fauna and flora remains have been found. Samples from 'Shustoke A' are about 2800 years old, and their fauna and flora indicate forest conditions. 'Shustoke B' samples were much more difficult to date, but radio-carbon analysis suggested 410 ± 90 years B.C., an age rather younger than was expected on geological grounds. Whatever its absolute age, the fauna of Shustoke B is predominantly one of ruderal herb community with both arable and pasture plants. A great variety of phytophagous species such as Chrysomelidae and Curculionidae, feeding mainly on such plants, was present and dung beetles such as *Aphodius* spp., though not especially numerous, were also present. A rather similar fauna was found in the lower part of a Roman well at Barnsley Park, Gloucestershire (Coope and Osborne, 1967), where the influence of man was further indicated by the presence of domestic refuse and remains of dwellings.

Historic times

The clearing of the wood continued through historical times, by means of the axe, fire and ring-barking. In each county, woodland was cleared to make way for crops, and trees were felled for fuel and building purposes (Darby, 1952). The woodland glades supported a higher density of livestock which in turn extended the open spaces and stopped woodland regeneration. The animals were much more effective than man in destroying root systems and preventing the growth of seeds and saplings. Little was done to halt this destruction, and even where individual trees were preserved or afforestation attempted, the trees often died through exposure in the otherwise open environment.

Most of the woodland was cleared in order to produce more grain. The arable fields had to be as extensive as possible in order to prevent food

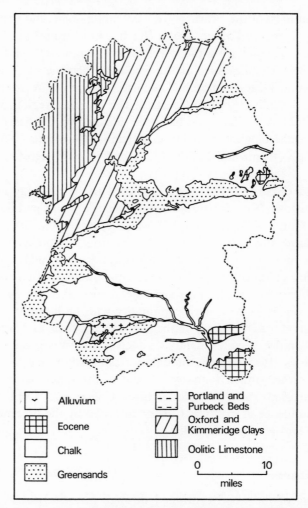

Alluvium		Portland and Purbeck Beds	
Eocene		Oxford and Kimmeridge Clays	
Chalk		Oolitic Limestone	
Greensands		0 10 miles	

Fig. 1.5a The geology of Wiltshire, for comparison with pattern of land use shown in Fig. 1.5b.

running out during the ensuing winter, spring and early summer, and to provide sufficient seed for the next planting season. The size of the harvest was the most important factor in the life of each community and family. Out of 321 harvests between 1480 and 1800, Hoskins (1964, 1968) has estimated that 26% were deficient, 34% were average, and 40% were good. The average harvest is defined as one where the average price was within 10% of the norm, calculated on the basis of a 31-year moving average. Deficient and good harvests were over and below 10% of the

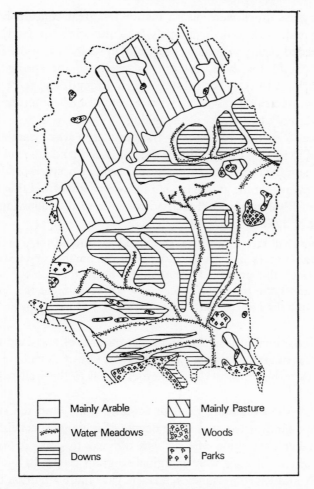

Legend:
- Mainly Arable
- Water Meadows
- Downs
- Mainly Pasture
- Woods
- Parks

Fig. 1.5b The pattern of land use in Wiltshire in the late eighteenth century, based on a map published in 1794. (Davis, 1794.)

norm, respectively. Since the yields rarely exceeded the ratio of 1:3·0 or 4·0 in the medieval period, a large proportion of seed had to be kept back for planting in the following year. A bad harvest almost automatically prejudiced the next harvest through a deficiency of seed, and this may have been the reason for series of bad harvests. 1506, for example, was the first really good harvest for 7 years, when the weather came to the rescue and yields were doubled so as to improve the balance between grain for seed and grain for human food.

But the grass crops were also a matter for great concern because the farmer depended on draught animals for cultivating the land, and these animals needed plenty of fodder, especially in spring when the farmer had quickly to build up their strength in readiness for spring ploughing (Simkhovitch, 1913; van Bath, 1963). Their strength had to be sustained; according to one calculation, the work accomplished by an ox fell by up to 50% after only two weeks of ploughing.

There was a further reason why the grass crop was essential. Manure was needed to restore the fertility of the cultivated land; the animals spent the daytime on the downland, grass heath or valley pastures, and the night in the fold manuring the arable land. This was so important that the tenant of a farm in Blythburgh, Suffolk, had to agree to the following covenant in his lease of 1771:

> The tenant shall keep at least 800 sheep and shall fould them regularly at all seasonable times in the year with the usual number of hurdles upon some part of the premises that shall be likely to be most benefitted thereby, under the penalty of £5 a night for every night the flock or at least one hundred sheep part thereof shall be foulded or lodged off the premises.[3]

The need for animal feed was so great that the area of grassland frequently exceeded that of arable by over 50%. Many tracts of downland and grass heath, far from being waste land, were essential to the well-being of the arable fields.

In spite of the overriding need to produce grain, and the consequent need for grasslands, there was some degree of regional specialization, and an even higher proportion of land was devoted to grass in some areas, especially where the soils were marginal to grain crops. In such areas as the Romney and Lincoln Marshes, farmers gave more attention to producing meat, milk and other animal products. Fig. 1.5 is based on a land-use map of Wiltshire, included in the Board of Agriculture report of 1794, which indicates the main areas of grassland and arable. The Oxford and Kimmeridge clay areas of north-west Wiltshire were primarily pasture land, and the streams and rivers in the chalkland parts of the county were bordered by flood- and water-meadows. Arable land was more in evidence on the Great Oolite formations of the Cotswolds and the chalklands of Salisbury Plain and the Marlborough Downs – except on the higher downlands where the drier soils were left to pasture. Farm output reflected these variations in topography and land use. Although both arable and grassland were present in all parts of the county, the claylands and alluvial soils were most famous for their cattle and such animal products as butter and cheese; the limestone and chalk soils were more important for grain (Davis, 1794).

The proportion of land devoted to grass varied not only from one region to another, but from one century to another. The most dramatic changes occurred when there was a shortage of human food. There is evidence that much marginal land was reclaimed for cultivation from the twelfth to the mid-fourteenth centuries in response to a rising population. This was the case at Mere, Wiltshire, where the general trend was reinforced by the development of a woollen industry and important market. The town was the fourth most important vill in the county in 1377, in terms of taxable population. In order to sustain the increased population, areas of woodland were cleared on the heavy low-lying Kimmeridge Clays, and the communal arable fields were also extended on to the near-by chalk scarp. The chalk soils were clearly marginal to cultivation because the demesne land 'upon the hills' in 1300 was valued at only 2 pence per a. compared with 8 pence per a. for 'under the hills', but the need for extra food was so pressing that the scarp was cultivated in the form of strip lynchets. Taylor (1966) believes that all the present-day lynchets, cut as ledges in the steep slope, were formed in this period.

The grassland–arable balance reflected the broad trends in both English and European trade. There was an extension of grasslands in the late fifteenth and early sixteenth centuries owing to a rise in the relative value of sheep husbandry. It is estimated that the average price of wool, between 1462 and 1486, was 29% higher than that for 1450 to 1461, whereas the average price of grain actually fell by 2%. Many landowners took advantage of the growing profitability of wool production by increasing the size of their flocks. They also benefited from a reduction in the cost of their overheads since wages comprised only 11% of the total expenses on a sheep holding compared with 36% on arable land. By extending the area of pasture land and thereby increasing the size of the flocks, sheepmasters made considerable economies of scale (Bowden, 1967). As the arable area shrank, many settlements diminished in size and population. The labourers of the former mixed farm economy left their homes, and over 1500 settlements were actually deserted (Beresford and Hurst, 1971). To quote one example, William Cope was granted the manor of Wormleighton, Warwickshire, by the Crown in 1498, and he immediately set about purchasing all the lands and tenements owned by other men in the manor, and by 1499 he had succeeded in converting 240 a. (97·2 ha) of arable land to enclosed pasture, causing 60 persons to leave the village (Thorpe, 1965). In a comparatively few cases, the site of the original settlement and its

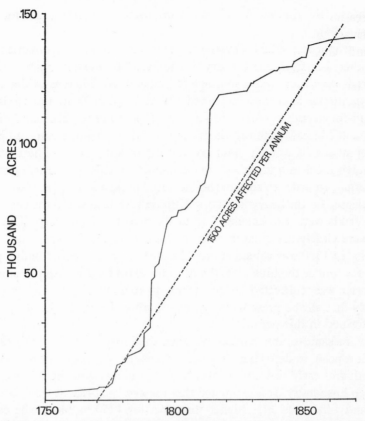

Fig. 1.6 The amount of rough grazing affected by enclosure in Somerset, in the eighteenth and nineteenth centuries. (*After* Williams, 1972.)

environs remain undisturbed under the grass. The medieval earthworks of Chalford, Oxfordshire, can still be recognized, and the grassland cover contains such species as woolly thistle (*Cirsium eriophorum*) and dwarf thistle (*C. acaulon*), which are usually indicative of old grassland.

From about 1750 onwards, the marked growth of industry and towns affected the use of agricultural land. Rowley (1972) has drawn attention to the migration of labour to such districts as the Shropshire coalfield, so that over half the population of Shropshire in 1801 was working in industry and living in towns, consequently dependent on purchasing grain, meat and milk from elsewhere. These developments encouraged landowners and occupiers to seek ways of increasing crop yields and extending the area of land under grain. The high prices of the Revolutionary and Napoleonic Wars between 1793 and 1815 provided a further incentive, and one method of increasing output was by sub-dividing extensive areas of common land into private holdings, often under the aegis of an Act of Parliament (Chambers and Mingay, 1966). This made it easier for the more progressive farmers to plough up the rough grazings for cultivation, and Williams (1972) has noted a significant increase in the area of rough grazings enclosed in Somerset between 1790 and 1820. Fig. 1.6 shows how enclosure was carried out in the county, slowly at first, and then with great rapidity. There were nevertheless regional variations in the diffusion of this form of land improvement – the process of enclosure began earlier in the Somerset Levels and Mendips, and later on Exmoor and the Blackdown Hills, reflecting perhaps the capability of the soil and agricultural prosperity in each area at that time (Table 1.3).

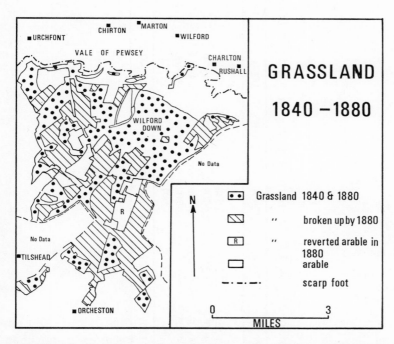

Fig. 1.7 Changes in the area of grassland on part of Salisbury Plain between 1840 and 1880, based on the Tithe Commutation surveys and the first large-scale survey of the Ordnance Survey.

Wheat prices were sufficiently high for it to be worth cultivating the more marginal soils of the Wiltshire downlands in the nineteenth century (Caird, 1852). Fig. 1.7 shows how landowners and occupiers extended the arable area in three of the parishes which extend from the Vale of Pewsey

Table 1.3 The area affected by the enclosure of rough grazings in each district of Somerset (as indicated by Williams, 1972)

Districts	1770s	1780s	1790s	1800s	1810s	1820s	1830s	1840s	1850s
					Acres *				
Levels	8 088	7 116	32 539	1804	7 811	–	1263	99	225
Mendips	2 100	3 331	12 509	1 685	4 470	–	–	237	217
South Hills	–	–	1 749	645	4 314	–	2357	2676	–
Western Hills	600	–	2 702	3699	22 400	786	2459	3445	4935
Other districts	398	–	601	77	946	–	116	57	–
Total	11 186	10 447	50 100	7910	39 941	786	6195	6514	5377

* 1 acre = 0·405 ha

southward on to the scarp and dip slope of Salisbury Plain. Many acres of downland were reclaimed by a process called paring and burning, whereby the turf was cut into pieces and pared from the ground, then piled up and burned, leaving ashes which were scattered and ploughed into the reclaimed surface. Many of these fields were given the name of Bakeland. As with many other husbandry practices, the short- and long-term benefits were keenly debated. Thomas Davis condemned the use of paring and burning on light soils (Stone, 1800) – although one or two bumper crops were taken:

> the consequences must be, that the roots which hold the land together and have been a century in forming, are destroyed all at once, and it is the work of another century to get a sward equal to the former; and, as arable land, it is totally incapable of bearing any crops after the first or second. This has been the case with thousands of acres of the best turf on the Wiltshire Downs.

The botanist was no less horrified. Claridge Druce (1886) learned how the soldier and monkey orchids were 'tolerably plentiful' on the chalk slopes around Whitchurch, Oxfordshire, until the late 1830s, when his informant was 'grieved and horrified to see the steep slopes pared and burnt in order to enrich the land with ashes, and . . . roasting alive' both kinds of orchid.

Many farmers hesitated before breaking up their grasslands on account of the cost and effort involved. According to Albert Pell (1899):

> while the field laughs with grain, it is more than possible that the owner groans at the cost of its artificial fertility . . . it would have been better to have left the down unbroken, the copse ungrubbed, the gorse and heather to bloom in peace, the sullen clay undrained.

There were many examples of costly failure: Joseph Stevens (1874) recalled seeing 'hundreds of acres of land' on Salisbury Plain, covered by small broken flints, where the once arable land had been allowed to revert to pasture owing to its low yields. In the light of this experience, farmers reacted very slowly and hesitantly to any rise in grain prices.

A high proportion of the newly won arable fields reverted to grassland during periods when wheat prices were relatively low. This was because farmers on the more marginal soils could not meet the high cost of their overheads at such times. Between 1874 and 1914, nearly 4 million a. (1·6 million ha) of arable reverted to grassland largely because of the import of relatively cheap grain from America. By the 1870s, railways had penetrated into the potentially rich grain-growing areas of America, and steamers had proved that they could maintain a reliable service across the north Atlantic. For the first time, it became possible to import large quantities of grain, grown in America, and prices fell below 50 shillings a quarter in 1875, below 40 shillings after 1883, and 30 shillings a quarter after 1893. English farmers on marginal soils could not compete, especially in poor harvests. Previously market prices had risen and offset the low yield when harvests were poor, but this ceased to happen after the 1870s because American farmers did not 'share' the bad weather, and continued to send large quantities of wheat at relatively low prices (Perry, 1973).

The reduced income from wheat sales affected the entire farm economy in some areas: landowners and occupiers had less capital to invest in land husbandry, new crops and techniques. Economy became the prime object-ive in farming and, according to Coppock (1962), this generally led to an extension of grassland. Many farmers reduced the size of their labour forces, which further encouraged the extension of grassland. It must be stressed that graziers did not always suffer to the same extent as 'cereal farmers'. Although imports of wool and meat depressed prices after about 1880, graziers were helped in two ways (Fletcher, 1961). First, the decline in grain prices meant that cattle-cake and other animal foodstuffs were cheaper. Secondly, many graziers were able to turn to liquid milk produc-tion – a commodity which could not be imported from abroad.

Dairy farmers were helped by a series of improvements in the distribu-tion and retail of milk. Until the 1860s, most of the milk for liquid con-sumption was produced by town dairies and those in the immediate vicinity of urban centres. Elsewhere most of the milk was used for butter and cheese, or fed to livestock, but, by the 1860s, a dense railway network had been established in lowland England and it became possible for the first time to transport milk quickly and cheaply to London and other major towns. Contracts were arranged between wholesalers and dairy farmers in such counties as Wiltshire and Essex, and by 1880 over 20 million gallons (90·0 million litres) were transported daily to urban outlets. By the early 1900s, the Great Northern Railway ran a special train from Burton-on-Trent to London, a distance of 150 miles, carrying milk from the farms of Derbyshire and Nottinghamshire. The Great Western Railway into Paddington became known as the Milky Way, and the pattern of land use was accordingly modified in the vicinity of the railway stations (Orwin and Whetham, 1971; Taylor, 1971). The establishment of the Milk Marketing Board in 1933 largely removed the differences between the prices paid for liquid milk and manufactured milk products, and the intro-duction of standard collection charges, irrespective of the remoteness of the farm holding from the dairies, reduced the importance of location factors (Baker, 1973). It encouraged the intensification and expansion of dairying wherever conditions were suitable for grass growth (Coppock, 1964).

The Government did not interfere in the balance of arable and grassland until 1917, when there were fears of an impending food famine. The Food Controller and the Board of Agriculture calculated that arable crops sustained four times as many people as animal products from the same area of land, and, as a result, the Board of Agriculture allocated a quota of grassland to be ploughed up in each county for cultivation during 1917–18. For the first time, the Board took powers to intervene on every farm, and direct the use of land (Sheail, 1973). The farmers of Huntingdonshire were ordered to plough up an aggregate of 20 000 a. (8094 ha) of land that had reverted to grass since 1874. It was hoped this land would be easier to reclaim and less valuable to the grazier than older grasslands but, in spite of the use of compulsory powers and the provision of auxiliary labour, horse teams and tractors by the Board and its county agents, the county War Agricultural Executive Committees, the goals were not met. Only 11 300 a. (4577 ha) of grassland were ploughed up in Huntingdonshire. £110 were spent in trying to reclaim 255 a. (103·3 ha) of rough grassland in Stilton Fen: prisoners of war were employed in clearing and renovating the drainage channels to facilitate ploughing, but the horse teams failed to make any headway, and the project had to be abandoned[4]. This type of grassland was simply not worth reclaiming, although it had been established for less than 40 years.

During the 1930s, the Government made renewed efforts to adjust the proportion of land under arable and grassland, through the use of subsidies, guaranteed 'farm-gate' prices and, to a lesser extent, through compulsory powers. The first grants were merely designed to slow down the decline in agricultural productivity by helping to correct such fundamental problems as bad drainage and soil acidity. The Agricultural Act of 1937 provided drainage grants and also gave financial incentives to farmers to apply lime and basic slag to grass and arable lands.

The renewed threat to food imports from submarine warfare caused the Government to give unprecedented support to large-scale reclamation schemes between 1939 and 1945 (Murray, 1955). In contrast to the 1914–18 war, there were attempts to reclaim not only the land which had reverted to grass during the years of depression, but also older established grasslands. The widespread and more efficient use of tractors and such machines as the gyrotiller made this possible.

The destruction of older grasslands threatened the rare and distinctive wild plant species associated with this habitat. Ecologists and the voluntary nature conservation societies were alarmed, and G. F. H. Smith, the chairman of the Wild Plant Conservation Board and the Honorary Secretary of the Society for the Promotion of Nature Reserves, made an urgent appeal to the Ministry of Agriculture in 1943, stressing:

> where land has never been cultivated or at least not since the critical days of the Napoleonic Wars it may be the habitat of rare British plants, and if it be ploughed up such plants may be exterminated to the grave detriment of botanical science.[5]

The Ministry accordingly instructed its county War Agricultural Executive Committtees to warn botanists whenever the requisition and destruction of grassland were contemplated, and the West Riding committee thereby gave warning of plans to break up and cultivate most of Farnham Mires. Dr G. Taylor succeeded in persuading the committee to 'spare' 25 a. (10·1 ha) of grassland where bird's-eye primrose (*Primula farinosa*) and meadow saffron (*Colchicum autumnale*) were found. He convinced the committee 'that it would be a catastrophe to disturb an area so rich and on which plants are still to be found in the same place as they were recorded 300 years ago'. But liaison often broke down or no agreement was possible. In May 1944, the Wild Plant Conservation Board appealed for the protection of some Hertfordshire meadows only to discover that the tractors had almost completed their work.[6]

Most observers regarded the losses in old grassland as a wartime phenomenon: Tansley (1945) commented:

> it is scarcely probable that the extension of agriculture will go much further, for the limits of profitable agricultural land must have been reached in most places.

But, after a pause, the destruction continued, encouraged by the Agricultural Acts of 1947 and 1957 which helped to maintain and increase the flow of capital into agriculture by enabling the Government to provide price guarantees for all major farm products, including cereals and fat stock. The guaranteed prices were determined each March after an annual review of the farm economy, and they have had the effect of encouraging farmers to plough up extensive areas of formerly permanent grassland for cereal production.

Although such legislation as the Town and Country Planning Act of 1947 awarded the Government and county councils unprecedented powers in the planning field, there were no powers to regulate changes in the use of farmland, and consequently many grasslands of interest to the biologist have been destroyed. Even some of the Sites of Special Scientific Interest, notified by the Nature Conservancy, have been affected. About 20 of the 190 scheduled sites in Dorset, Hampshire and Wiltshire have been so badly damaged by ploughing that the scheduled areas have had to be reduced. In Dorset, the area of 11 chalk grassland sites declined by an aggregate amount of almost 20% during the 1960s.

Today the surviving areas of old grassland are generally small and fragmented. The Ordnance Survey identified 59 separate grass heaths in

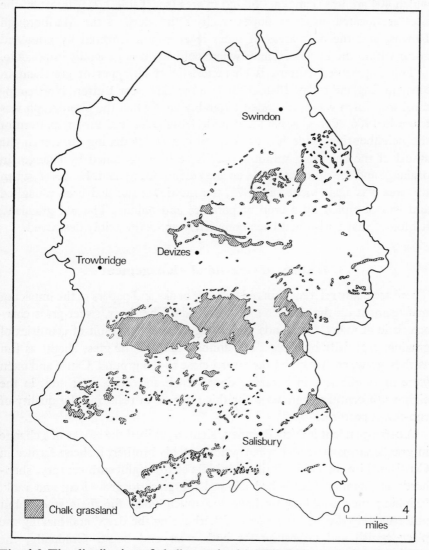

Fig. 1.8 The distribution of chalk grassland in Wiltshire, as recorded in 1966.

the Breckland of Norfolk and Suffolk in the 1880s, with an average area of 941 a. (381 ha): today the number of grass heaths has fallen to 37, with an average area of only 88 a. (35·6 ha) (Duffey, in press). This reflects the extension of cultivation, the creation of a large forest by the Forestry Commission, and the construction of airfields. Fig. 1.8 shows the fragmented pattern of chalk grassland in Wiltshire: the 76 130 a. (30 833 ha) of grassland in the county are distributed between 529 separate sites, of which 404 are less than 5 a. (2·0 ha) in area (see Table 1.2). Almost all the sites are located on steep slopes, such as the scarp of the Marlborough Downs, and the only areas of more level ground covered by grassland occur within the Defence Lands where cultivation is generally impossible.

The fragmented pattern is characteristic of all types of grassland in lowland England today. Plate 3 shows a meadow near Lutton, Northamptonshire, which was surrounded by arable land when the photograph was taken in 1966. The grassland and underlying ridge and furrow pattern of the neighbouring fields has been destroyed by ploughing, and even the sward of the surviving meadow may have been modified by changes in drainage and the nutrient status on the neighbouring land. However, when the area was again visited in 1971, the meadow itself had been ploughed and was occupied by a crop of potatoes and onions. The old grassland habitat of that parish had been completely and irreversibly destroyed.

History of Grassland Management

There are many references in old farming books and papers to the impact of management on the character of grasslands. The short, sweet grass characteristic of many downlands was thought to be the result of centuries of grazing, especially by sheep. The animals nibbled the grass shoots as fast as they grew, and cropped the sward close to the ground. Cattle and oxen were more effective in breaking down scrub and deep litter and, in the eighteenth century, were used for this purpose in restoring the quality of run-down pastures.

A correspondent in New Romney, Kent, described the effects of a change in grassland management when writing to his brother, a sheep farmer in Cardiganshire, in 1786. Until the middle of the eighteenth century, shepherds in Romney Marsh had allowed the grass to grow long and rank, 'thinking they could never have too much grass for their stock', but unfortunately this had the effect of 'harbouring the dews' and making the place aguish and unhealthy for both man and beast. Gradually, the shepherds discovered that sheep were healthier and fattened well on shorter grass. According to the correspondent, the results were startling: the shepherds

> keep more stock upon the land and feed it as close to the ground as they can without starving their sheep whereby the land carried more stock; the ground can harbour no dews, and the country is become much more healthy (Skilbeck, 1956).

The character of grasslands was also affected by wild animals, and most especially by the rabbit. Since about 1750, rabbits have had a considerable impact on the structure and floristic composition of grasslands, especially those of the chalk, limestone and sandy soils. The animal is not a native to Britain and is thought to have been introduced soon after the Norman Conquest (Veale, 1957). The animals were kept in warrens for their meat and fur, especially on well-drained sites around the coast and in lowland England. Very few rabbits lived outside the warrens until the eighteenth century, when the feral population began to rise dramatically and, by 1900, there were colonies of rabbits in most parts of lowland England (Sheail, 1971b).

Ant-hills were a serious nuisance on many grasslands: a report for Rutland in 1794 recommended spreading the contents of the hills over the sward on light soils, as a valuable top dressing, but on heavy soils the farmers were advised to take the debris away. In some Rutland meadows, the sward was so broken with ant-hills that graziers had to plough up the meadows and plant a fresh mixture of grass seed (Crutchley, 1794).

It was possible for a grassland site to remain undisturbed for many years by ploughing and afforestation, and yet change completely in structure and composition due to a cessation of grazing or mowing. Sheep were traditionally ubiquitous, and Coppock (1964) has estimated that there were over 50 animals per 100 a. (40 ha) in most arable areas in the 1870s. According to the Annual Returns, there were 17 552 000 sheep in 1971 compared with about 20 237 000 in 1871 (Table 1.4), but these figures conceal a fundamental change that has taken place in the distribution and character of the flocks. The number of animals has increased on upland holdings, as is suggested by the returns for Wales and, at a larger scale, for Caernarvonshire, where Hughes et al. (1973) have recently demonstrated their impact on the vegetation of the Snowdonia range. But sheep are no longer ubiquitous, and 74% of the holdings in England and Wales at the present day are without sheep, especially those farms in lowland areas. This is suggested by the returns for England and, at a larger scale for Suffolk, where the size of the flock has fallen from about 430 000 in 1871 to 43 000 in 1971 (Table 1.4). Lakenheath Warren in Suffolk, for example, was traditionally grazed by about 2000 sheep but the figure fell to 600 during the 1940s, and grazing ceased altogether in 1956 (Crompton, 1972). The age-composition of the flocks has also changed, with a significant increase in the proportion of animals under one year of age in all areas.

Table 1.4   Changes in the number and character of the sheep flock, as recorded in the Annual Returns of 1871 and 1971

		Sheep	
	Number Thousands	1 year and over	Under 1 year Percentage
England and Wales			
1871	20 237	65	35
1971	17 552	53	47
England			
1871	17 530	64	36
1971	11 499	51	49
Wales			
1871	2 706	70	30
1971	6 053	57	43
Suffolk			
1871	430	56	44
1971	43	48	52
Caernarvonshire			
1871	216	72	28
1971	488	60	40

Changes in the distribution and character of the flocks are symptomatic of profound changes in the farming economy. The number of animals fell during the cereal depressions of the late nineteenth century and inter-war period owing to the relatively high cost of folding the sheep on arable land. The expansion of the dairy industry caused some lowland farmers to sell or reduce the size of their flocks in order to secure more food for their cattle. The sale of meat was affected by competition from imports, especially from New Zealand, and the Government encouraged dairying at the expense of meat and wool production during both wars. Finally, the housewife has come to prefer smaller joints, which are produced more easily by hill breeds of sheep. These economic and social changes have had the effect of removing the dominant biotic factor from many grasslands, although the effect was obscured in some parts because of the concurrent rise in the number of rabbits until myxomatosis broke out in the mid-1950s (see page 113). The former sheep pastures soon became invaded by scrub.

The spread of scrub did not always indicate a fall in the intensity of management. A tract of scrub was carefully conserved in the parish of Ellesborough, Buckinghamshire, as a source of fuel. When the arable and grasslands of the parish were enclosed in 1805, special provision was made so that the poor people of the near-by village could continue to 'cut, take and use Scrubbs, Furze, Fern, or other Fuel'[7]. Scrub was also used as game cover and, in some cases scrub species were actually planted: farmers on the grass heaths of East Suffolk planted parts of their sheep-walks with grass seed, mixed with furze and broom, during the early nineteenth century[8].

There was sometimes a clear-cut distinction between grasslands set aside for grazing and those which were used for hay. Three meadows near Oxford are an extreme case: Port Meadow was probably the area described as common pasture in the Domesday Book of 1086 (Darby and Campbell, 1962), and there is no evidence that the site has been mown, except during the Civil War when hay was cut for the King's horses between 1643 and 1645. In contrast, Pixey and Yarnton Meads have been regularly cut, and the aftermath has been grazed (Gretton, 1912). The botanical composition of the meadows has reflected these differences in management. Baker (1937) listed a total of 95 species in the three meadows, of which 56 occurred in the grazed and 69 species in the two hay meadows; 26 species were confined to the grazed meadow and 39 to the hay meadows. There has been a relatively long history of uniform management because of the involved legal status of the sites and the large number of people who had the right to depasture animals and take hay (Farrer, 1936), and this has had the effect of stopping individual farmers from ploughing up the grasslands, either for cultivation or re-seeding.

The grazing and hay rights, which were owned by the same group of people on Yarnton and Pixey Meads, were owned by different groups of men on the areas of grassland called Lammas Lands. North Meadow, Cricklade in Wiltshire, is a fine example of a Lammas Land. The grass is subdivided into allotments, and each farmer is responsible for harvesting the hay in his allotment. Fig. 1.9 shows how the hay rights were allocated in 1970 on North Meadow. Over the years, there has been some amalgamation reflecting the increase in farm size and the reduced importance of hay as winter fodder. Livestock are excluded from the meadow until the 12th August each year when the hay is removed and the animals are admitted to graze the aftermath. The livestock of each commoner are allowed to graze the entire meadow, irrespective of how the hay rights have been awarded. This pattern of land management has tended to inhibit any changes or 'improvements' in grass husbandry, including the application

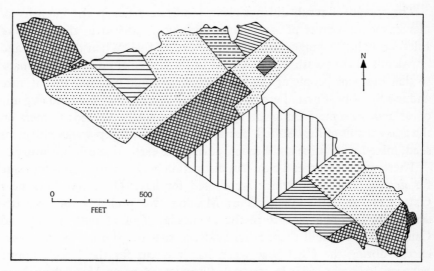

Fig. 1.9 The distribution of hay rights in North Meadow, Cricklade, Wiltshire. The allotments of each of the 10 commoners are represented by a different shading.

of fertilizer or herbicide, and this means that North Meadow, for example, contains the largest colony of fritillary (*Fritillaria meleagris*) in Britain, together with its associated flora. In order to perpetuate this form of management and thereby protect the wild species, the Nature Conservancy has recently acquired most of the hay allotments and, in March 1973, the site was declared a National Nature Reserve.

The ratio of pastoral to arable land was clearly reflected in the pattern of management, especially of the communal grasslands. At Wigston Magna, Leicestershire, each person was forbidden by the manorial court from depasturing more than eight cows and 40 sheep for every yardland, or approximately 30 a. (12·2 ha), held in the common arable field. But during the sixteenth and seventeenth centuries, the population of the village more than doubled, and the arable lands had to be extended in order to sustain the greater population. This encouraged an increase in the density of animals grazing the common grasslands, leading to overgrazing and a greater risk of the animals becoming weak through hunger and more vulnerable to disease. The outbreak of serious epidemics among cattle in 1711–14 and 1744–45 in many parts of the country may have been symptomatic of this trend toward overgrazing (van Bath, 1963). In the face of these difficulties, each person clearly had to reduce his personal stint of grassland in the communal grounds and, at Wigston Magna, this was formalized by a decree from the Court of Chancery which reduced the stint to four cows and 24 sheep per yardland (Hoskins, 1957).

Some of the oldest grasslands were associated with badly drained conditions. They included coastal marshes, fenlands and flood-meadows, whenever the near-by river was in spate. In general, farmers welcomed these floods: the salt left by the high spring tides on the Suffolk marshes was said to protect sheep from the rot and diseases.[9] The river waters were frequently warmer than the air and land in winter and early spring, and often rich in nutrients: the grass in the flood-meadows grew up to three weeks earlier in spring, thereby helping to sustain the ewes at lambing time when there was little alternative food available.

However, farmers could not fully exploit these grasslands until the floodwaters were regulated. The jurors of Gedney, Lincolnshire, in 1607, complained that a high spring tide could flood the marshes to a depth of 6 ft (1·83 m) in 3 hours and drown all the livestock (Thirsk, 1965). At Houghton, on the river Ouse in Huntingdonshire, the risk of flooding was

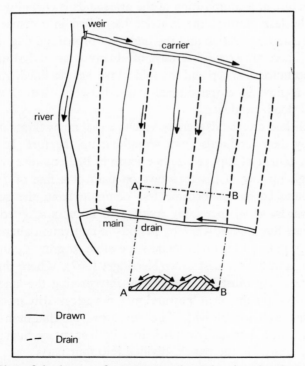

Fig. 1.10 Plan of the layout of a water-meadow, showing the channels used for drainage and irrigation.

so serious that men were ordered to help clear the meadows of livestock whenever they heard the church bells giving warning, by 'ringing backwards'.[10]

During the late medieval period, farmers discovered ways of both draining and irrigating the grasslands of the flood-meadow, through the construction of hatches and sluices across the streams, and carriers and drains through the meadows. Fig. 1.10 indicates how the water was directed on to the grasslands by the system of carriers and allowed to flow as a moving sheet of water through the grass into the drains in the furrows. The water-meadows were generally irrigated in February and grazed by sheep in March and April, followed by further irrigation for a hay crop in July. The meadows were then set aside for aftermath grazing. This kind of management programme created a grass-dominant meadow, with few dicotyledons present (Sheail, 1971a). It produced the kind of habitat shown in Plate 4 of the Breamore meadows on the river Avon in Hampshire (see page 5). The photograph was taken in January 1934, and Plate 5 shows the same view in March 1970. The meadows have not been irrigated since 1949 and the structure and composition of the grasslands is changing. Since the middle of the last century, the practice has fallen into disuse owing to changes in farm organization and the high cost of maintaining the drawns and drains. There are now other ways of obtaining an early bite of grass, and the importance of sheep and the foldcourse has declined. As a result, the habitat is now in a transitional stage, between that of the water-meadow and flood-meadow.

Old-established grasslands were so valuable that many farm tenants were prevented by their landlords from ploughing up the turf. In 1682, for example, the tenant of Upper Cainhoe Farm in Bedfordshire had to agree not to plough up any of his meadows on pain of a fine of £5 per a.[11] However, these covenants became less common from the seventeenth century onwards, as an increasing number of farmers adopted a system called alternate husbandry, whereby periods of cultivation alternated with periods of ley grass. This often made it possible to grow grain and root crops on the land for the first time (Kerridge, 1967). Where this practice occurred, great care must be exercised in interpreting the land use data given in surveys, for the term 'pasture' may not necessarily mean that the grassland was well established. The surveyors of Hampton-in-Arden, Warwickshire, were unusually explicit in a Parliamentary Survey of 1649 in describing a close in the parish, divided into four parts:

> part thereof is att present plowed, and the rest of it unplowed, it being a usuall course with the Inhabitants to plow their grounds which they doe call

pasture for two or three yeares together and then to lett it lye for pasture fifteen or twenty yeares and then plowe it againe, and this they doe with a great part of their pasture ground.[12]

This change in land husbandry was facilitated by the more widespread use of vetches and field beans, and by the introduction of new fodder crops. Clover seed was imported from the Low Countries as early as 1620, although many years elapsed before it was grown in many parts of the country. It was first grown in the Vale of Pewsey, Wiltshire, in the 1690s, when there was a dispute over the tithes for grass and clover at Great Cheverell.[13] Clover and sainfoin were not grown in the Yorkshire Wolds until the 1740s, and did not become widespread until after 1780 (Harris, 1961). Turnips were sown in parts of Norfolk and Suffolk during the mid-seventeenth century, but there are no records of the crop in the Yorkshire Wolds until a century later. These fodder crops were grown in rotation with cereals. Farmers on Salisbury Plain and the Hampshire Downs in the middle of the nineteenth century adopted the rotation wheat/vetches and turnips/barley/clover, which sustained their livestock and reduced their dependence on old grassland (Little, 1845). In addition, clover, sainfoin and lucerne helped to maintain, and in some cases improve, the nutrient status of the soils, because of the ability of bacteria in their root systems to fix nitrogen. Since nitrogen is one of the most effective elements in promoting plant growth, the crops were extremely important in this context, further reducing the farmer's reliance on farmyard manure in sustaining the fertility of the soil.

At the same time, a search was being made for grass species, able to grow in the 'dead season' of the year as part of a crop rotation (Fussell, 1964). Guides were published to the various types of grass, and the Society for the Encouragement of Arts and individual persons encouraged the commercial production and sale of clean grass seed. Beddows (1969) has illustrated the difficulties of popularizing such native species as timothy (*Phleum pratense*) and the cocksfoot (*Dactylis glomerata*). There was confusion over the vegetative characteristics of timothy, and farmers found it difficult to distinguish the flowering heads from those of meadow foxtail (*Alopecurus pratensis*). The seeds were often planted at the wrong time of the year, on badly prepared seed beds, and farmers were soon disheartened by the poor results.

Until the nineteenth century, the production cycle on a mixed farm holding had been almost closed: it produced wheat, barley, meat and some wool for sale, and hay, roots, clovers and rotation grasses for consumption by the livestock which manured the land and pulled the plough and which, in turn, ensured larger grain yields. This closed circuit was broken when farmers began to buy larger quantities of feed-stuffs and fertilizers. Oil-cake was used as a manure and increasingly as an animal foodstuff, together with maize which was imported on a large scale from the 1840s onwards. Bone dust was used on chalk soils by the 1840s for sustaining the fertility of arable land and, by 1850, superphosphate of lime was being produced from a mixture of bones and sulphuric acid, and became the most popular artificial fertilizer for turnips in Wiltshire. Thompson (1968) has estimated that the amount of feed and fertilizer used on farms doubled between 1864 and 1877, and rose by 10 and 27 times respectively over the period 1830 to 1880. They gave the farmer greater flexibility in planning his cropping programmes, and further reduced his dependence on old grasslands.

A way was found of increasing the productivity of those areas which remained under grass. The adoption of the Thomas process of steel-making in the 1880s produced a phosphatic fertilizer called basic slag. This was especially useful in improving poor pastures. The fertilizer was relatively cheap, easy to apply, and experiments conducted by Gilchrist and others at Cockle Park, Northumberland, after 1896 demonstrated how it promoted the growth of white clover, which had previously been a small component of poor swards (Pawson, 1960). The effect of this increased clover growth was twofold: it provided a more nutritious fodder and the nitrogen-fixing properties of the legume enhanced grass growth.

In spite of all these improvements, many fields remained under grass owing to the uncertainties of drainage. The claylands of Buckinghamshire were described by one writer as 'a puddle in winter, and cracked and dried as hard as a brick in summer'. In some cases, farmers tried to drain the surface by constructing a series of ridges and furrows across the meadows – the water ran down the sides of the ridges into the furrows which characteristically contained rushes (*Juncus* spp.) and coarse grasses. From the seventeenth century onwards, farmers experimented with various forms of under-drainage (Darby, 1964). Tile drains were widely used after 1800, and, by the 1840s, ways had been found of manufacturing large numbers of cylindrical tiles at very low prices. Under-drainage was primarily a means of improving poor arable land, especially on heavy soils, but Phillips (1972) has recently drawn attention to efforts to drain grasslands. On the estate of the Earls of Scarborough in the West Riding of Yorkshire, over 20% of the drained area was pasture. There were experiments in mole-draining in the eighteenth century, but the practice did not become significant until the 1850s when John Fowler invented the mole plough, which was adapted for use with a steam engine.

In view of the high cost of drainage work, there was little activity during periods when cereal prices were low. The years of the two world wars saw great improvements. Under-drainage was not only extended, but schemes which had fallen into decay through lack of maintenance were brought back into use. The technology of land drainage made great progress during the 1939–45 war and the subsequent years. According to Trist (1970), under-drainage has been the most important improvement to occur in British agriculture over the last 30 years. It has allowed many farmers to break up their hitherto-permanent grasslands. Even where this has not taken place, the structure and character of the grasslands have changed, reflecting the changes that have occurred in the water regime, both on and below the surface.

These developments have reduced the value of old-established grassland to such an extent that they are an anachronism on many holdings. As long ago as the sixteenth century, such commentators as Anthony Fitzherbert (1523) advocated the ploughing up of grasslands, full of moss, and the cultivation of the land for a few years: he believed the re-seeded grassland would benefit from these improvements for up to 10 years. As a refinement, Barnaby Googe (1577) recommended that the final crop should be oats, which were a 'great breeder of grass', together with the seed swept out of the hayloft.

Following the introduction of scientifically balanced seed-mixtures and the improvements in fertilizers and under-drainage, the value of ley grasslands rose dramatically and, in 1937, the Council of the Royal Agricultural Society of England recommended the annulment of all covenants forbidding the ploughing up of old grasslands. The covenants were ended by Parliament in May 1939, and a grant of £2 per acre was made available to those farmers who ploughed up grasslands of over seven years in age and over 2 a. in extent. Within a few months, Sir George Stapledon, the leading advocate of ley farming at that time, had published a book giving data on ploughing techniques, seed-mixtures and the management of leys (Stapledon, 1939).

The official advocacy of rotational grasslands gave impetus to the destruction of many of the remaining botanically rich meadows. In 1938–39 one of the last remaining fritillary meadows at Mickfield, Suffolk, was roughly ploughed and drainage work had begun. In addition to the fritillary, the meadow contained the green-winged orchid (*Orchis morio*), common spotted-orchid (*Dactylorchis fuchsii*), dyer's greenweed (*Genista*

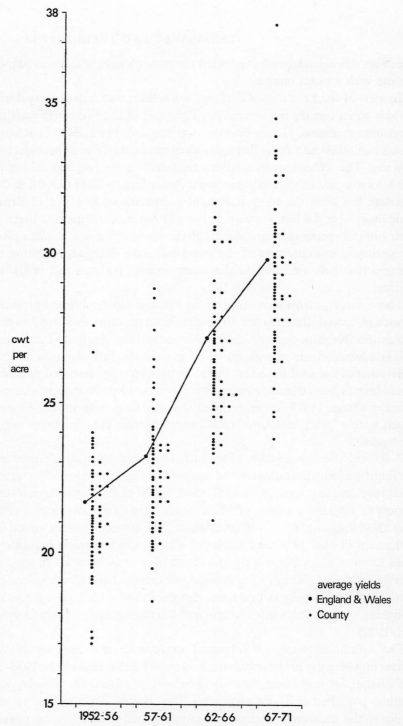

Fig. 1.11 Changes in the estimated yield of hay from permanent grassland in England and Wales and for each county.

tinctoria), creeping jenny or money wool (*Lysimachia nummularia*), common twayblade (*Listera ovata*), and the adder's tongue fern (*Ophioglossum vulgatum*). A local naturalist warned the Society for the Promotion of Nature Reserves that 'the countryside was rapidly becoming less floriferous in this mechanical and destructive age' and that in time every ancient meadow would disappear unless a conscious effort was made to save a sample. In the event, an appeal was launched for the acquisition of the Mickfield Meadow, and the site was eventually purchased and given to the SPNR for safekeeping. Soon afterwards, all the adjacent meadows were ploughed and their distinctive flora became locally extinct[14].

Not only has the area of old grasslands diminished, but the character of the remaining sward has changed. This may be indicated by the rise in the estimated output of hay from permanent grassland, as published by the Ministry of Agriculture each year. Fig. 1.11 shows how the yields of the five-year periods have risen over the last 20 years in England and Wales. The changes coincide with the increased application of fertilizers, herbicides, and the installation or improvement of existing under-drainage. According to Church and Webber (1971), nitrogen was the fertilizer most widely applied to grasslands, and Fig. 1.12 indicates its increased use over the period 1952 to 1971.

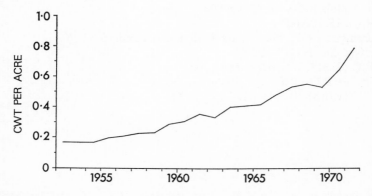

Fig. 1.12 The increase in the application of nitrogen (cwt per a.) on agricultural land in England and Wales between 1952 and 1972. (*Data analysed and provided by* F. H. W. Green.)

The impact of these changes on wild plants and animals may be illustrated by an example from the County of Huntingdon and Peterborough. According to the Annual Returns of 1971, there were 33 442 a. (13 544 ha) of permanent grassland and rough grazings in the county. A botanical survey was carried out on these grasslands in order to assess their floristic interest, and only about 880 a. (356 ha) were found to be of interest – that is, only about 3–4% of the area recorded in the Annual Returns. The composition of the remaining grasslands had been affected by the application of fertilizers and herbicides and, in some cases, by earlier cultivation. The grasslands of botanical interest were distributed between 42 sites, of which 40 were less than 40 a. (16·2 ha) in size. 41% of the grasslands fell within the boundaries of two sites, separated by the river Ouse.

This kind of botanical survey needs to be repeated in other counties, but clearly profound changes are taking place in the wildlife of grasslands, and their conservation is a high priority, otherwise the old grassland habitat is likely to disappear altogether in many parts of the countryside.

Manuscript sources

1 Buckinghamshire Record Office (R.O.), IR 63A.
2 Public Record Office (P.R.O.), IR 29/30 31/42.
3 East Suffolk R.O., 50/22/1.11.
4 Huntingdon R.O., War Agricultural Executive Committee, minute books, Huntingdon.
5 Society for the Promotion of Nature Reserves (S.P.N.R.) MSS, Box 24, 1/75.
6 S.P.N.R. MSS, Box 24 1/83–87.
7 Buckinghamshire R.O., IR M 21/1.
8 East Suffolk R.O., HA 61 436/892.
9 P.R.O., IR 18/9683.
10 Huntingdon R.O., Manchester MSS ddm 6, bundle 5.
11 Bedfordshire R.O., L 4/306.
12 P.R.O., E 317, Warwickshire 12.
13 Wiltshire R.O., 155
14 S.P.N.R. MSS, Box 15 1.1.

2 Classification of grassland communities in Britain

There has been much argument and discussion among plant ecologists in recent years on the classification of plant communities. Part of the discussion has centred on what are basically technical points – how the data should be collected, size of sampling unit, distribution of samples and whether the community should be characterized by dominant species, characteristic species, or even by groups of species. This issue has been further confused or clarified, according to one's point of view, by the development of mathematical techniques, such as association analysis, nodal analysis and others which use modern computers to sort species into groups which can be used as units in a system of classification.

The second, and more fundamental, point of dispute among different schools of ecology, which has been responsible for the production of several systems of classification, is the question of existence and nature of the plant community.

The main problem in the classification of grasslands is the same as in most plant communities – is the variation continuous or discontinuous? Do the sharp boundaries which we use to separate different groups of plants really exist, or are they generalizations drawn from an amalgam of ideas and field data?

Whatever the solution may be, most people are agreed that classification is necessary for two main reasons: first, to produce some order into a collection of facts; secondly, to enable us to communicate descriptions and ideas on relationships about the type of vegetation recorded, and make comparisons with similar or dissimilar samples from elsewhere. (For a fuller discussion of some of these points, the reader should see Professor D. A. Webb's stimulating essay 'Is the classification of plant communities either possible or desirable?' in *Bot. Tidskr.* **51**, 362–370).

The classification used in this book is a development of the system originally proposed by Sir Arthur Tansley. It is based on the belief that most grassland workers prefer a system which enables them to name a grassland community in fairly precise terms without adopting a rigid hierarchical system which pre-supposes that the units are necessarily related.

A further reason for not adopting the phytosociological* system used extensively on the Continent is that not all grassland types in Britain have been analysed in this way. In some cases British workers have created new 'associations' or else different species have been used to characterize vegetation units. Although these variations in methodology are generally understood by botanists, cross-reference to continental phytosociological names will be made in the following account for the convenience of the reader.

Grasslands in Britain can be divided into three main groups depending on soil type: calcareous, neutral and acidic, and each of these is considered below in detail.

Calcareous Grasslands

Calcareous grasslands are distributed over a variety of geological formations which are largely composed of calcium carbonate. The limestones vary lithologically, but the main ones are Cretaceous chalk, Oolitic (Jurassic) limestone, and Carboniferous or Mountain limestone. Calcareous grasslands are also found on limestones with large amounts of magnesium, and on a variety of calcareous igneous and metamorphic rocks. Related communities occur in Breckland on soils derived from the chalky boulder clays and there are also close relationships with communities of stable calcareous sand dunes on the coast, notably the 'machair' of the Outer Hebrides.

The soils derived from these different parent materials are relatively similar chemically (Table 2.1) and it is this common factor which gives the characteristic composition, both in terms of floristics and structure, to calcareous grasslands. These soils are characterized by a high pH (usually in the range 5·5–8·4), a high available calcium ion content (usually in the range 300–1000 mg Ca/100 gm), a high free calcium carbonate content (30–75%) and often a high organic matter content (7–20%). The soils

* Phytosociology is the science that deals with the detection of regularly recurring combinations of plant species, their naming and classification and their correlation with the more readily observed environmental factors.

Site	Parent rock material	pH	Loss on ignition %	Total nitrogen %	K mg/100 gm	Ca mg/100 gm	Mg mg/100 gm	P mg/100 gm	CaCO$_3$ %
Aldbury Nowers, Herts	Chalk	7·6	12·8	0·74	17	820	13	2·34	71
Aston Rowant, Oxon (1)	Chalk	7·6	21·8	0·87	12	1510	18	1·20	32
Aston Rowant, Oxon (2)	Chalk	7·7	18·3	0·84	14	1460	19	2·30	37
Aston Upthorpe, Berks	Chalk	7·4	22·0	0·74	17	830	17	1·64	66
Barton Hills, Beds	Chalk	7·3	11·3	0·49	22	660	13	1·39	46
Crown Point, Dorset	Chalk	7·4	31·4	1·44	30	1390	24	1·90	41
Devil's Dyke, Cambs	Chalk	7·6	11·0	0·32	18	430	10	0·98	69
Hambledon Hill, Dorset	Chalk	7·6	18·1	0·93	15	1100	15	0·70	60
Porton Down, Wilts	Chalk	7·8	26·0	0·50	11	1300	16	3·70	63
Barnack Hills and Holes, Northampton	Oolitic limestone	7·1	11·5	0·35	18	440	8	1·09	78
Honnington Camp, Lincs	Oolitic limestone	7·6	15·4	0·47	31	550	15	0·96	34
Shacklewell Hollow, Rutland	Oolitic limestone	7·8	7·5	0·30	12	320	7	0·65	91
Black Head, Co. Clare*	Carboniferous limestone	5·8	19·5	n.d.	23	1560	n.d.	n.d.	Medium
Brean Down, Somerset*	Carboniferous limestone	6·9	4·0	n.d.	19	360	n.d.	n.d.	Medium
Cote Gill, Arncliffe*	Carboniferous limestone	7·9	18·7	n.d.	14	2320	n.d.	n.d.	Medium
Crook Peak, Somerset*	Carboniferous limestone	6·8	14·2	n.d.	12	2080	n.d.	n.d.	High
Purn Hill, Somerset*	Carboniferous limestone	6·4	23·4	n.d.	41	1520	n.d.	n.d.	Low
Anstey's Cove, Devon*	Devonian limestone	7·4	12·3	n.d.	62	1280	n.d.	n.d.	Medium
Cronkley Fell, W. Yorks*	Sugar limestone	6·5	14·6	n.d.	15	1680	n.d.	n.d.	High
Breidden Hill, Montgomery	Dolomite	5·2	22·9	n.d.	50	360	n.d.	n.d.	Low

* Source: Proctor (1956). Extractant N/2 ammonium nitrate, EEL. flame photometer. Organic carbon by Walkley and Black method. pH by glass electrode. Other values by method given in Wells and Barling (1971).
n.d. = no data

43

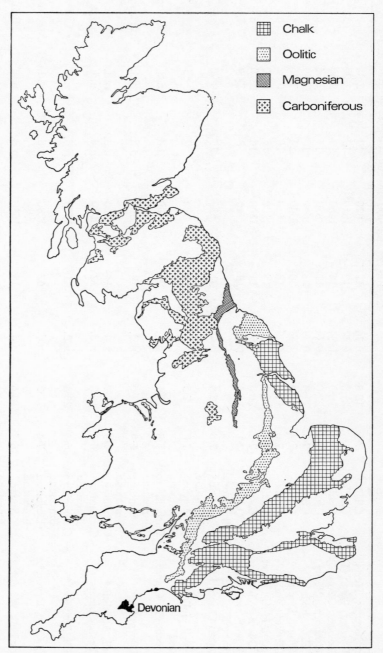

Fig. 2.1 The distribution of the major calcareous rock formations in England, Wales and Scotland.

are often of the rendzina or brown earth type. They tend to be porous and therefore dry, but where there is clay or compacted drift derived from calcareous rocks, drainage may be impeded and give a transition from dry grassland to marsh or even mire.

Although rocks and soils are mainly responsible for the many types of grassland ranging from calcareous to non-calcareous, climatic and historical factors interacting with aspect differences determine, for the most part, the variation in botanical composition.

Lowland calcareous grasslands may be sub-divided according to the major geological formations. That is, Chalk, Oolitic limestone, Carboniferous limestone, Devonian limestone, Magnesian limestone and other calcareous rocks. The distribution of the major formations is shown in Fig. 2.1.

Despite their wide geographical variation in floristics, lowland calcareous grasslands have a sufficiently large number of constant vascular plant species to give them a degree of homogeneity, and, in order to measure the similarities and differences, botanists use a concept defined by the term *constancy*. This is the repeated occurrence of certain species in different samples of grassland. It is conventional to distinguish five classes of constancy, which in reverse order are: Class V constants occurring in 81% or more of the samples, IV in 61–80% of the samples, III 41–60%, II 40–21%, I in less than 20%. The Class V constants of calcareous grassland in Britain are: *Briza media, Festuca ovina, F. rubra, Carex flacca, Lotus corniculatus, Plantago lanceolata, Poterium sanguisorba* and *Thymus drucei*. Other species with a high constancy throughout the range of grasslands but which do not always reach the value of Class V constancy in all calcareous grasslands are: *Carex caryophyllea, Koeleria cristata, Helictotrichon pubescens, H. pratense, Agrostis stolonifera, Trisetum flavescens, Orchis mascula, Leontodon hispidus, Scabiosa columbaria, Campanula rotundifolia, Helianthemum chamaecistus, Linum catharticum, Veronica chamaedrys, Ranunculus bulbosus* and *Polygala vulgaris*.

All of the dry, base rich, anthropomorphic grasslands in Britain are included in the order Brometalia erecti Br.–Bl. 1936 in the classification used by phytosociologists (Shimwell, 1971a). With the exception of a few calcareous grasslands developed on the coast, which are placed by Shimwell in the sub-alliance Xerobromion Br.–Bl. and Moore 1938, the remainder of the lowland calcareous grasslands are referable to the sub-alliance Eu-Mesobromion Oberd. 57 (Shimwell, 1971a). Upland calcareous grasslands are assigned by Shimwell (1971b) to another sub-alliance, the Seslerio–Mesobromion Oberd. 57.

Chalk grassland

Chalk deposits laid down during the Cretaceous period are widespread in western Europe (Fig. 2.2) and are particularly abundant in France and England. The soils which develop from the underlying chalk rock are shallow rendzinas and they support plant communities which are among the richest and most interesting in Europe. Stott (1970) provides a useful historical review of the development of studies (which began in 1827) of chalk grasslands in France. No useful purpose would be served by listing the names used by French botanists of the many variants of this vegetation and the reader should consult Rose (1965) and Bournérias (1968) who discuss much of the information available at the alliance level. Rose

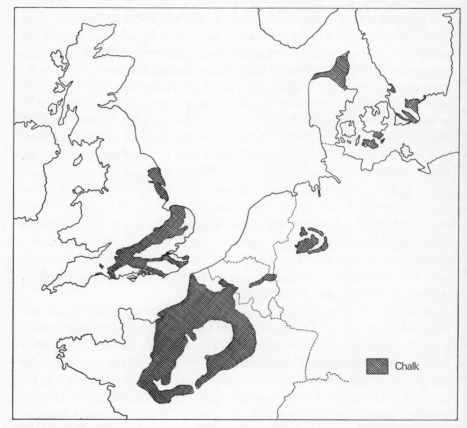

Fig. 2.2 Distribution of chalk deposits in western Europe. (*From* Shimwell, 1973.)

(1972) looks in some detail at the phytogeographical relationship between the floras of the French and English chalk.

In Britain the chalk outcrop extends from Beer Head, Devon in the south-west on a north-east axis across Dorset, Wiltshire, Hampshire, Berkshire and the Chiltern Hills to the Yorkshire Wolds, with easterly extensions forming the North and South Downs. Much of the area which is shown as chalk on the maps of solid geology is covered with drift and superficial deposits which completely hide the effect of the underlying rock. Many of the downs formerly used as sheep walks have now been reclaimed for arable farming, and old chalk grassland is more or less restricted to steep scarp slopes which are unsuitable for ploughing. An exception to this generalized statement is the Salisbury Plain area, where grassland still survives on gentle slopes which are included in the Ministry of Defence's training area.

The classical studies of Tansley (1922), Tansley and Adamson (1925, 1926) provide descriptions of chalk grasslands in Sussex and Hampshire and other studies by Hope-Simpson (1941) and Cornish (1954) add to our knowledge of chalk grasslands in southern England but there has been no adequate description yet of the full range of floristic variation encountered on English chalk grasslands. The following seven main types of chalk grassland were recognized by us as a result of extensive surveys made from 1965–70 throughout the whole range of chalk grasslands. They are classified according to differences in the dominance of grass or sedge species, namely:

Festuca ovina/rubra   *Arrhenatherum elatius*
Carex humilis   *Helictotrichon pubescens*
Zerna erecta   Mixed Gramineae
Brachypodium pinnatum

The *Festuca* type has long been the most widespread of all chalk grassland communities and it may be that the *Zerna erecta* and *Brachypodium pinnatum* types which are spreading represent stages in the retrogression of chalk grassland following cessation of grazing. This has been brought about both by the decline in the numbers of sheep in lowland England and by the elimination of rabbits after myxomatosis. *Festuca ovina* is the main colonizer of bare chalk surfaces but as the sere progresses *Festuca rubra* is able to enter the grassland, probably as a result of a build up of organic matter and nutrients, and most closed chalk communities consist of a mixture of these grasses. When grazed or mown, *Festuca* swards form the high quality chalk swards which have long been associated with sheep

walks, as many as 40–45 species of flowering plants and mosses per m²
being recorded.

Carex humilis grasslands, which are only found in any quantity in Wiltshire and Dorset, are unusual in two respects. First, they contain a large
number of herbs, grasses and mosses which are able to grow together without any one achieving dominance, even under a variety of grazing pressures,
and secondly, the presence of *Carex humilis* in the moister, western grasslands and its absence from the drier Kentish chalk. In Europe, this species
has a continental type of distribution and is a component of the dry steppe-
like *Stipa* grasslands, and its anomalous position in England has yet to be
explained.

Grasslands in which *Zerna erecta* or *Brachypodium pinnatum* are dominants occur widely in the Yorkshire Wolds, Kent and parts of the Chiltern
Hills but are less common elsewhere, although locally it is not unusual to
find areas where these species are abundant while the remainder of the
grasslands are of the *Festuca* type. Both types of grassland can maintain
their floristic richness if managed correctly but, if not grazed, burnt or
cut, other species are rapidly eliminated by competition.

The *Arrhenatherum elatius* type of grassland is often associated with
some past disturbance factor, such as ploughing or re-seeding. The *H.
pubescens* type is characteristic of moist north-facing aspects, with an
abundance of mosses but with few dicotyledons. The factors responsible
for the Mixed Gramineae type have yet to be fully elucidated.

Oolitic (Jurassic) limestone grassland

These grasslands are best seen in the Cotswolds in Gloucestershire where
they occupy thousands of hectares on steep, rocky slopes, often on 'common land'. Three main types are recognized, dominated by *Festuca ovina*,
Brachypodium pinnatum and *Zerna erecta*, the last two species often being
intermixed. Floristic variation within these three groups is similar to that
in the corresponding types described for the chalk. The *Festuca* type is the
richest grassland sward on limestone and includes species such as *Helianthemum chamaecistus*, *Poterium sanguisorba* and *Cirsium acaulon* which
may be more abundant than on the chalk. Many of the grasslands are
managed by annual burning, a practice not common on the chalk and which
may be responsible for the predominance of *Brachypodium pinnatum* on
the oolite. Rare species such as *Anemone pulsatilla* (Plate 6) are particularly
characteristic of oolitic soils but only two species, *Thlaspi perfoliatum* and
Stachys germanica, occur in these grasslands and not on the chalk.

Carboniferous limestone grassland

This ecological formation has a more westerly and northerly distribution than the chalk and oolite, occurring in regions with a higher rainfall. The major outcrops are in northern England, North Wales and Ireland, with scattered outliers in the south, for example the Mendip Hills.

Festuca ovina/rubra swards are again the most common type of grassland association but many southern species, such as *Zerna erecta* and *Brachypodium pinnatum* which reach their climatic limit on the northern limestones, are poorly represented or absent. The high rainfall in these regions, together with the increased occurrence of drift deposits, and the uneven and rocky nature of the terrain, gives great diversity in edaphic and microclimatic conditions and it is not uncommon to find acidiphilous species such as *Potentilla erecta* growing a few centimetres from a strict calcicole such as *Scabiosa columbaria*. Species more characteristic of acid grasslands, such as *Sieglingia decumbens* and *Carex pulicaris*, become increasingly common in these grasslands, and the greater spectrum of mosses clearly reflects the moister climate.

Plant communities on the Carboniferous limestone have been studied intensively in recent years by Ivimey-Cook and Proctor (1966), Williams and Varley (1967), Shimwell (1971b) and Lloyd (1972). All but the last-named author used phytosociological methods to classify the vegetation. Their results illustrate clearly some of the difficulties for, despite the fact that in some instances the same communities were sampled, the names given to them vary, depending on individual judgement. This not only applied to abstract units at the nodum level, but also to the sub-alliance level, higher in the hierarchical classification. For example, Shimwell (1971b) places the *Dryas octopetala* communities of the Burren, Co. Clare in the sub-alliance Seslerio-Mesobromion Oberd. 57, while Ivimey-Cook and Proctor (1966) place the same community in the sub-alliance Mesobromion erecti Br.-Bl. and Moore 38. It is beyond the scope of this book to describe the complex grassland communities found on the Carboniferous limestone and only a bare outline of the range of variation is given below.

On steep-sided, humid valleys in the Peak District of Derbyshire and in the Mendip Hills, the association Helictotricho-Caricetum flaccae is common, characterized by the co-dominance of *H. pratense* and *Carex flacca*, in association with a variety of calcicolous herbs. Another group of grasslands characterized by the dominance of *Sesleria albicans* forms a well-marked zone across northern England and western Ireland, and is placed by Shimwell (1971b) in the sub-alliance Seslerio–Mesobromion Oberd. 57. Differential species in this group include *Galium sterneri*, *Polygala amara*, *Dryas octopetala* and *Epipactis atrorubens*.

Lloyd (1972) used association analysis and ordination methods to classify Carboniferous limestone grasslands in the Sheffield region. He recognized seven types of grassland which can be divided into two major groups. The first group contains *Arrhenatherum elatius* as a constant, usually with *Festuca rubra* and other mesophilous grasses. Some of these communities, especially those on the deeper brown-earth loams containing mixed herb communities, clearly fall into the Arrhenatheretalia of the continental botanists. The second group, which includes species-poor and species-rich facies of the *Festuca ovina* type, are clearly Mesobromion grasslands, closely allied to the southern chalk grasslands.

In Ireland, much of the underlying Carboniferous limestone is covered with drift or peat deposits which completely mask the calcareous nature of the rock, but in the Burren region the limestone is exposed and forms one of the most dramatic landscapes in the British Isles. Ivimey-Cook and Proctor (1966) place the grasslands developed over the limestone pavement in the sub-alliance Mesobromion erecti Br.-Bl. and Moore 38, recognizing three noda and one association. The *Dryas octopetala-Hypericum pulchrum* association is the commonest grassland community, occurring on dark organic soils overlying the bare limestone, and is characterized by the constancy of *Dryas octopetala* and *Calluna vulgaris* with a large group of calcicolous mosses, including *Breutelia chrysocoma*, *Hylocomium brevirostre* and *Neckera crispa*. The presence of *Dryas*, an arctic–alpine species, at sea-level is a surprising feature of these grasslands.

Transitions from the *Festuca* and *Sesleria* type of grassland to a sub-montane type occur at altitudes above 1000 ft in which northern and upland species such as *Antennaria dioica*, *Saxifraga hypnoides*, *Potentilla crantzii*, *Galium boreale* and *Minuartia verna* occur. This forms a distinct sub-group within the 'normal' Carboniferous limestone type.

Magnesian limestone grassland

This formation extends as a narrow strip northwards from Nottingham to Durham and is entirely lowland in distribution. Most of the semi-natural grasslands on this formation have been destroyed by mining activities but the small areas still extant form an important link between the southern chalk and oolitic grasslands and those of the northern Carboniferous limestone. In general, floristics are similar to the oolite, including grasslands in which *Zerna erecta* and *Brachypodium pinnatum* are dominant. Nevertheless, they occur far enough north to have northern elements, such as *Sesleria albicans* and *Epipactis atrorubens*, growing with southern species, for example *Inula conyza* and *Blackstonia perfoliata*. Such features are of special interest to the ecologist.

Shimwell (1971b) places the grasslands on the Magnesian limestone escarpment in Durham which contain *Sesleria albicans* in the association Seslerio–Helictotrichetum. He considers this to be the most thermophilous association in the alliance. On north- and west-facing slopes, where the microclimate is more humid, two variants of the Caricetosum pulicariae occur which contain first *Selaginella selaginoides* and *Pinguicula vulgaris*, and secondly *Antennaria dioica*. These have many species in common with the Seslerio–Caricetum pulicariae, which is widespread on the limestone uplands of the northern Pennines from the Craven district of North Yorkshire to the Teesdale fells.

Devonian limestone

This formation outcrops in a very limited area in the Torquay–Brixham district of Devon and farther south on Plymouth Hoe. Most of the outcrops are on rocky cliffs and although these grasslands show floristic affinities with the southern chalk and limestone, the mild oceanic climate probably accounts for the number of rare and distinctive plants which occur, such as *Trinia glauca*, *Bupleurum baldense* and *Ononis reclinata*. The position of these grasslands in the phytosociological classification is far from clear. *Helianthemum apenninum* and *Trinia glauca*, which are well established at Berry Head and on the Carboniferous limestone at Brean Down, are used as character species of the Xerobromion. However, the paucity of other species of this type and the abundance of Mesobromion character-species suggests that for the present these maritime communities are best considered as intermediates and are not assigned to either alliance.

Other calcareous grasslands

Various types of sandstone, shale, greywacke, schist, gneiss and igneous rock locally contain sufficient calcium carbonate to give rise to soils which are calcareous or at least base-saturated. When situated in lowland areas

Table 2.2 Systematic arrangement of European meadow communities (*taken from* Williams, 1968)

Class Molinio–Arrhenatheretea Tx. 37

Order 1 Molinietalia Koch 26

  Alliance A   Calthion Tx. 36 (Bromion racemosi Tx. 51, Juncion acutiflori Br.–Bl. 47, in part)

    Association

    (a) Juncetum subnodulosi Koch 26

    (b) Senecio–Juncetum acutiflori Br.–Bl. Tx. 52

    (c) Deschampsia–Brometum (racemosi) Oberd. 57

    (d) Achilleo–Brometum (racemosi) Oberd. 57 (Holcetum lanati Issler 36)

    (e) Silao–Brometum (racemosi) Oberd. 57 (Sanguisorbo–Silaetum Klapp 51)

    (f) Polygono–Brometum (racemosi) Oberd. 57 (Angelico–Cirsetum Klapp 51, Bistortae–Brometum ⎫ (Bromo–
        Oberd. 56) ⎬ Senecionetum
          ⎭ Tx. 51)

    (g) Polygono–Cirsietum oleracei Tx. 51

    (h) Thalictro–Cirsietum oleracei Pass. 55

    (i) Chaerophyllo–Ranunculetum acontifolii Oberd. 52

    *(j) Trollio–Juncetum subnodulosi (Koch 26, Vollm 47) Oberd. 57

    *(k) Crepido–Juncetum acutiflori (Br.–Bl. 15) Oberd. 57

    *(l) Juncetum filiformis Tx. 37 (Filiformi–Scirpetum [Tx. 37] Oberd. 57)

    *(m) Polygono–Scirpetum (Schwick. 44) Oberd. 57

    *(n) Trollio–Cirsetum (Kuhn 37) Oberd. 57 (Cirsio–Valerianetum Kuhn 37, Cirsietum rivularis Ralski 31)

  Alliance B   Filipendulo–Petasition Br.–Bl. 47

    *Association Filipendulo–Geranietum Koch 26

  Alliance C   Molinion Koch 26

    Association

    (a) Molinietum Koch 26

    (b) Polygonetum bistortae Kovacevic 59

Order 2 Deschampsietalia caespitosae Horv. 56

   Alliance A Deschampsion caespitosae Horv. 35
     Association Deschampsietum caespitosae Horv. 30
   Alliance B Alopecurion pratensis (Agrostion albae Sóo [33] 40)
     Association
       (a) Alopecuretum pratensis Eggler
       (b) Galio–Alopecuretum Hundt. 58

Order 3 Arrhenatheretalia Pawl. 28

   Alliance A Arrhenatherion (Br.–Bl. 25) Koch 26
     Association
       (a) Arrhenatheretum medioeuropaeum (Br.–Bl. 19) Oberd. 52
       (b) Arrhenatheretum subatlanticum Tx. (37) 55
      *(c) Arrhenatheretum montanum Oberd. 52 (Arrhenatheretum holcetosum Oberd. 38, Centaureo nigrae–
             Arrhenatheretum Oberd. 57)
      *(d) Trifolio–Festucetum rubrae Oberd. 57
      *(e) Poa–Trisetetum (Knapp 51) Oberd. 57
      *(f) Melandrio–Arrhenatheretum (Kuhn 37) Oberd. 57

  *Alliance B Polygono–Trisetion Br.–Bl. 47
     Association
      *(a) Trisetetum hercynicum Tx. (37) 55
      *(b) Astrantio–Trisetetum Knapp 52
      *(c) Geranio–Trisetetum Knapp 51

   Alliance C *Poion alpinae (Gams 36) Oberd. 50
   Alliance D Cynosurion Tx. 47

53

* Predominantly montane associations

these frequently bear grasslands similar in floristic composition to those already described. The most important examples are found in Scotland, on such formations as the Silurian greywackes and shales in the Southern Uplands, but lowland examples are also to be seen in places such as Stannor Rocks in Radnorshire and Bredon Hill in Worcestershire.

Certain localized maritime grasslands, for example those on the machairs of the Hebrides and the Serpentine outcrops in Cornwall, are important types which have floristic similarities with lowland calcareous grasslands but it is beyond the scope of this book to deal with them in detail.

Neutral Grasslands

Tansley (1939) used the term neutral grasslands to include 'semi-natural grasslands whose soils are not markedly alkaline, nor very acid and mostly developed on clays and loams'. Although this is rather a loose definition covering a wide range of grasslands it is retained here because of its wide use and acceptance in the past. It is particularly useful because it clearly separates this important group of grass communities from those of calcareous and acid soils.

In large parts of central Europe neutral grasslands form an important part of the agricultural economy. Many are old and have been managed by the same methods for generations. Because of their agricultural importance, neutral grasslands have been intensively studied on the Continent. Ellenberg (1963) in particular gives a detailed account of their ecology in central Germany, while other workers, for example Oberdorfer (1970) and Westhoff and den Held (1969), provide floristic lists and discuss their systematic position in a European classification. Williams (1968), working with Professor Ellenberg at Zürich, made an outstanding contribution to our understanding of ecological processes in meadows by discussing and reviewing the extensive literature on these types of grasslands.

All neutral grasslands, both pastures and meadows, may be referred to the class Molinio–Arrhenatheretea R. Tx. 37. Originally this class was divided into two orders, the drier group forming part of the order Arrhenatheretalia, the wetter grasslands the Molinietalia. More recently, a third order has been created, the Deschampsietalia caespitosae Horv. 56, which accommodates those grasslands which do not fit comfortably into the other two orders.

The classification of European meadows used by Williams (1968) is shown in Table 2.2. It is clear from an inspection of this classification that some associations are not found in Britain, for example the Polygono–

Cirsietum oleracei Tx. 51, while others such as the Silao–Brometum racemosi Oberd. 57 are represented. The alliance Cynosurion Tx. 47 is widespread throughout Ireland (O'Sullivan, 1965) and is also common in lowland Britain. Nevertheless since there has been no systematic study of neutral grasslands in Britain using phytosociological methods, it is not possible to place them accurately in the hierarchy of European grasslands, and the system adopted in this book (p. 58) is based on the constancy of groups of species to well-defined grassland habitats.

Neutral grasslands are found throughout Britain, usually associated with calcareous geological formations, either derived directly from the underlying rock or from drift or alluvial deposits. They are usually situated below about 300 m O.D. and the majority are in the south and east but there is also an important group found at higher altitudes in northern England and Wales.

Water regime, management, soil and geographical location are the four main factors which determine the botanical composition of neutral grasslands.

Water regime

The composition of neutral grasslands is greatly influenced by the tolerance of individual species to different degrees of inundation. Thus *Sanguisorba officinalis*, a common plant of flood-meadows, is absent from areas which are under water for prolonged periods, for example the East Anglian Washes. On the other hand, *Primula veris* is unable to withstand any period of flooding so that although it occurs in many permanent meadows it is absent from flood-meadows.

Species characteristic of neutral grassland fall into groups which reflect their ability to withstand increasing periods under water:

Dry   1. *Festuca rubra, Anthoxanthum odoratum, Primula veris, Orchis morio.*

Moist 2. *Holcus lanatus, Alopecurus pratensis, Sanguisorba officinalis, Cardamine pratensis.*

        3. *Carex panicea, C. disticha, Juncus articulatus, Dactylorchis incarnata, Lychnis flos-cuculi.*

Wet   4. *Carex acutiformis, Juncus effusus, Caltha palustris, Galium palustre.*

        5. *Phragmites communis, Glyceria maxima.*

Management

The ecological interest of neutral grassland is often closely related to management, both past and present. In some cases, the land has been used in the same way for centuries. The famous fattening pastures of the Midland counties, for example, have been farmed by traditional methods from generation to generation, and the management of Lammas Lands has been prescribed by local by-laws. These sites are of special interest to the ecologist for they enable him to study the relationship between the development of grassland communities and particular forms of management.

The two principal forms of management are grazing or cutting for hay, and these will be considered in detail in Chapters 8 and 9. Unfortunately, from the conservationist's point of view, many of these grasslands are fast disappearing, especially in lowland England, as a result of modern agricultural techniques.

Soil

Dry calcareous grassland develops directly on soils derived from the parent rock, but most meadow grasslands occur on material deposited either by glaciation or by river action (alluviums). The physical and chemical properties of these soils (Table 2.3) depend on the nature of the transported material and on the processes of deposition. Alluvial meadows and water-meadows (see below) are in general on loams, whilst permanent meadows are on clays.

Different species within the same genus may show preferences for particular soil types within the meadow complex. *Juncus inflexus* occurs on heavy clay soils, while *J. effusus* shows a preference for peaty ground. *J. conglomeratus* is most common on soils with a high silt status. *Carex disticha* is common on soils with a high humus content, while *C. otrubae* prefers heavy clay soils.

There are also floristic differences associated with the variation in base status of neutral grassland soil. Thus, *Conopodium majus*, *Carex nigra*, *Filipendula ulmaria*, *Lychnis flos-cuculi*, *Betonica officinalis*, *Meum athamanticum*, *Centaurea nigra* and *Arrhenatherum elatius* occur on meadow soils of low base status, whereas others, for example, *Orchis morio*, *Orchis mascula*, *Listera ovata*, *Poterium sanguisorba*, *Primula veris*, *P. farinosa*, *Silaum silaus*, *Filipendula vulgaris*, *Carex flacca* and *Briza media*, are confined to

Table 2.3 Chemical analytical data of some neutral grassland soils*

Site	Grassland type	pH	Loss on ignition %	Total nitrogen %	K mg/100 gm	Ca mg/100 gm	Mg mg/100 gm	P mg/100 gm
Drostre Bank, Brecon	Neutral *Molinia*	6·1	37·9	1·86	24·0	753	20	2·7
Glyn Llech, Brecon	Neutral *Molinia*	6·1	77·0	2·76	34·0	1390	33	3·6
North Duffield Carrs, Yorks	Washland	7·1	7·0	0·69	5·4	331	10	0·6
Wheldrake Ings, Yorks	Washland	4·8	28·7	1·0	27·0	777	38	n.d.
Cantley, Yorks	Wet Alluvial	4·8	64·0	2·93	22·0	399	27	3·7
North Stoke, Sussex	Wet Alluvial	5·8	25·4	0·96	26·0	1020	30	n.d.
North Meadow, Cricklade, Wilts	Flood Meadows	6·5	29·7	1·20	19·0	1620	23	n.d.
Mickfield, Suffolk	Flood Meadows	6·0	10·1	0·34	9·3	467	18	n.d.
Langford, Dorset	Water Madows	6·2	66·2	2·80	20·0	1530	16	3·90
Downton, Wilts	Water Meadows	6·2	49·1	2·70	20·0	2220	18	4·80
Earls Croombe, Worcs	Ridge and Furrow	5·7	10·4	0·48	13·0	310	67	0·95
Corby Pasture, Lincs	Ridge and Furrow	6·1	12·1	0·45	11·0	545	31	0·46
Orton, Westmorland	Dales Hay Meadows	7·9	9·4	0·30	4·2	1010	14	n.d.
Gowk Bank, Cumberland	Dales Hay Meadows	6·3	44·1	1·50	36·0	1740	50	n.d.
Crosby Gill, Westmorland	Northern Grazed Meadow	5·5	34·7	1·53	23·0	1170	55	2·90

* Unpublished data of D. A. Wells
n.d. ... no data

the more base-rich soils. Many other meadow species are tolerant of a wide range of soil conditions e.g. *Succisa pratensis, Lotus corniculatus, Carex panicea.*

Geographical location

Climate and geology form the major factors affecting geographical location. Base-rich neutral grasslands in the north are on a relatively shallow calcareous drift soil derived from either the Carboniferous limestone or a closely allied calciferous sandstone. These conditions are absent in the south of England where this type of grassland is found in flood plains on deeper alluvial soils derived from the oolitic limestone or chalk. Typical southern species of these base-rich grasslands, like *Fritillaria meleagris, Orchis morio, Colchicum autumnale* and *Sanguisorba officinalis*, are also characteristic of those areas which are managed for hay-making.

Semi-natural grasslands on the calcareous boulder clays are confined to southern England and most have been destroyed by ploughing. This is particularly so in eastern England, where the local *Trifolium ochroleucon* meadows are almost a thing of the past. The base-poor type of wetter neutral grassland is found mainly in the west as illustrated, for example, by the distribution of *Carum verticillatum*. The species of *Dactylorchis* are characteristic of the corresponding base-rich areas, *D. praetermissa* occurring in the south, *D. purpurella* in the north.

Classification

Neutral grasslands in Britain have never before been classified using botanical criteria and the scheme presented below is an attempt to fill this gap. The classification, devised by D. A. Wells, is based on the constancy of groups of species to well-defined grassland habitats, together with distinctive species which separate one group from another.

Because neutral grasslands have been formed as a result of agricultural exploitation over the centuries they have acquired an agricultural nomenclature which has been retained where applicable for each group described below.

Neutral *Molinia* grassland

A dwarf grass/sedge community with *Carex nigra, C. panicea, Anthoxanthum odoratum* and *Holcus lanatus* dominant, with *Molinia caerulea* occasional to frequent. It is developed on alluvial soils in which the water table is at or near the soil surface for much of the year, but is dry enough in summer for grazing by cattle. *Dactylorchis praetermissa* is present in this group, with many other herbs.

Washlands

This is a distinctive, but relatively infrequent group found in Cambridge-shire and Huntingdonshire, in areas between rivers which are liable to prolonged winter flooding. The dominant species is *Glyceria maxima* in association with *Phalaris arundinacea*. The lower the water table the greater is the contribution of *Phalaris*.

The Washes are usually summer grazed by cattle and occasionally cut for hay. They are botanically poor, but most important as winter feeding grounds for migratory ducks.

Wet alluvial meadows

The grasses *Glyceria fluitans* and *Alopecurus geniculatus* are co-dominant, with *Carex hirta* and *Polygonum amphibium* well represented. Unlike the previous group, wet alluvial meadows are easily improved agriculturally by drainage, being replaced by an *Alopecurus pratensis/Cynosurus cristatus/ Holcus lanatus* community, or more frequently completely destroyed by conversion to arable agriculture. *Oenanthe fistulosa, Polygonum hydropiper, P. minus, Ranunculus sceleratus,* and *Rumex conglomeratus* form a distinctive group within this grassland type. They are used as summer grazings by cattle and occasionally cut for hay. *Rumex* spp. and *Polygonum* spp. provide seed for wintering wildfowl and the summer grazings provide suitable nesting conditions for many species of waders and other ground-nesting birds.

Flood-meadows

Areas of permanent grass beside rivers which flood for short periods in the winter are found in the Upper Thames basin, the South Midlands and in parts of East Anglia. These grasslands are nearly always cut for hay and only rarely is the aftermath grazed. Because of their situation they can usually be assumed to have had this form of management for many centuries with the result that a species-rich polydominant community has developed (Plate 7). Characteristically they contain a group of grasses of which *Alopecurus pratensis, Briza media, Agrostis stolonifera, Anthoxanthum odoratum*, *Cynosurus cristatus*, *Festuca pratensis*, *F. rubra*, *Holcus lanatus*, *Lolium perenne* and *Poa trivialis* are all occasional to frequent with a wide selection of herbs, the most common being *Filipendula ulmaria*, *Sanguisorba officinalis*, *Silaum silaus*, *Thalictrum flavum* and *Ophioglossum vulgatum*, (Plate 8).

These meadows contain some of the most beautiful and showy species in the British flora, *Fritillaria meleagris* (Plate 9) and *Leocojum aestivum* being perhaps the best known. The finest examples of this grassland type occur in the Lammas Lands along the river Thames.

Water-meadows

From about 1500 onwards, many flood-meadows were transformed into water-meadows, where the water regime was regulated by a system of weirs, artificial courses and earthworks. A time-table of irrigation and drainage was devised to provide an early 'bite' of grass for sheep and this is explained more fully in Chapter 1.

These meadows are dominated by grasses, especially *Festuca arundinacea*, *F. pratensis* and *Bromus commutatus*, because of selective weeding of dicotyledons. Of special note is the presence in these meadows of various hybrids between *Lolium perenne* and *Festuca pratensis*, confined to southern England and the Midlands.

Ridge and furrow old pasture

This complex group of grasslands is characteristic of heavy clay soils with impeded drainage and usually low fertility. It is distinguished by the wide variety of species present. Plate 3 shows a site of this type at Lutton, Northamptonshire.

The pattern of ridge and furrow was sometimes created by ploughing within the confines of the strips of the open-fields. Once formed, the ridges were preserved as they facilitated surface run-off. In other cases, the ridges were formed specifically for drainage, either of arable or grassland. These grassland sites today may be cut for hay and grazed in summer: many are grazed at low-stocking densities throughout the year.

They are characterized by a large number of grass species, similar to those listed for flood-meadows, but in addition with *Helictotrichon pubescens*, *Briza media*, *Phleum pratense* and a great variety of herbs, of which *Betonica officinalis*, *Primula veris*, *Rhinanthus minor*, *Centaurea nigra*, *Genista tinctoria* and *Serratula tinctoria* are characteristic.

Of special interest is the abundance of *Orchis morio* in some of these meadows and, in others, *Colchicum autumnale*.

Pastures on light soils, derived from Calcareous sandstones

Agrostis stolonifera, Anthoxanthum odoratum, Cynosurus cristatus, Dactylis glomerata, Festuca rubra and *Poa pratensis* form the basis of this group with many species which are also found in chalk grasslands. Modern agriculture has taken full advantage of these easily worked, free-draining soils and only fragments of this type of grassland are known. *Saxifraga granulata* can be used to differentiate a group characteristic of base-rich soils, and *Conopodium majus* an acid group.

Dales hay-meadows

This distinctive group of hay-meadows is found on the Carboniferous limestone or on other calcareous substrata at altitudes of between 700 ft to 1000 ft O.D. (213–305 m) in the north of England and in central Wales. These grasslands have been cut for hay for many centuries and at fairly regular intervals receive dressings of farmyard manure, but, until recently, no artificial fertilizers.

They are characterized by a long list of species of which the most common are: *Alchemilla* spp., *Crepis paludosa, Geum rivale, Sanguisorba officinalis, Trisetum flavescens, Trollius europaeus* and *Cirsium heterophyllum*. Of the grasses, the most frequent are: *Anthoxanthum odoratum, Bromus mollis, Dactylis glomerata, Festuca rubra, Holcus lanatus, Molinia caerulea* and *Poa trivialis*.

Regional differences exist, *Platanthera chlorantha* and *Tragopogon pratensis* being found in the central Wales hay-meadows, while *Cirsium heterophyllum* and *Geranium sylvaticum* are more characteristic of the English hay-meadows.

Northern grazed meadows

Unlike the southern grazed meadows, in which botanical composition is clearly related to the grazing factor, the northern grazed meadows owe their composition to a combination of soil–water factors and grazing. The mineral-rich flushes in which this type of grassland occurs are generally small, but nevertheless distinct from the surrounding grassland.

They are characterized by *Alchemilla* spp., *Briza media, Carex panicea, Lathyrus montanus, Pinguicula vulgaris* and *Valeriana dioica*. Other species which are not so common but usually present and which assist in identifying this group are *Bartsia alpina, Parnassia palustris, Primula farinosa, Serratula tinctoria* and *Sesleria albicans*.

Tussocky neutral grassland

Deschampsia cespitosa, Juncus inflexus, Lysimachia nummularia and *Ranunculus ficaria* form the nucleus of this type which occurs on heavy soils with impeded surface drainage, most often used for summer grazings by cattle. A closely related and structurally similar group occurs on peaty, lighter soils with a high water table but not with impeded drainage and is characterized by *Juncus effusus, Holcus lanatus, Poa trivialis, Ranunculus acris* and *Stellaria graminea*.

Reverted pasture

This, the most widespread of neutral grasslands, develops by 'weed species' invading sown pastures or as 'tumble down' grassland following arable. Sporadic attempts to improve production from some of the grasslands already described, by drainage, fertilizers and herbicides, often result in the formation of this kind of grassland. Characteristic species are: *Cynosurus cristatus, Dactylis glomerata, Lolium perenne, Trifolium repens, Agrostis* spp., *Bellis perennis, Centaurea nigra, Cirsium arvense* and *Ranunculus acris*. This type of grassland clearly belongs to the alliance Cynosurion cristati, which is usually split into the association Lolio–Cynosuretum and Centaureo–Cynosuretum.

The fattening pastures of north Northumberland, Leicestershire and the Romney Marsh form part of this group.

Acidic Grasslands

Acidic grasslands are the most widespread type of semi-natural grassland in Britain. They occur on a great variety of soil types (Table 2.4), from the free-draining acid sands of the Dorset heaths to upland and montane soils of Scotland and Wales. Many of these montane soils vary considerably, from water-logged peaty soils with a moder humus, to relatively free-draining mineral soils with a mull humus. The common factor linking this

Table 2.4 Chemical analytical data of some acidic grassland soils

Site	pH	Loss on ignition %	Na mg/100 gm	K mg/100 gm	Ca mg/100 gm	P mg/100 gm
†Corrie Fee, Clova, Angus	5·2	10·9	3·2	16·0	135·5	1·05
†Craig Formal, Perthshire	5·4	22·1	11·7	21·9	204·5	0·19
†Near Loch Chuir, Perthshire	6·0	11·9	5·2	7·3	226·2	0·03
†Meall an Fheur Loch, Sutherland	4·6	10·4	5·0	18·4	24·0	0·42
†Beinn Bhreac, Inverness-shire	5·3	12·8	5·7	13·4	75·8	0·44
†Clune Burn, Inverness	5·7	7·1	6·1	6·9	57·0	0·28
†Harris, Isle of Rhum	6·2	27·8	44·0	13·9	102·1	0·13
†Ben Wyvis, E. Ross	5·2	17·5	4·7	11·2	39·5	0·37
†Glen Doll, Clova, Angus	5·4	13·0	5·6	13·4	76·0	0·44
*Maes y Gaer, Aber, Caerns	5·1	10·2	n.d.	8·3	6·9	0·31
*Nant Anafon, Caerns	4·9	8·3	n.d.	7·5	5·9	0·04
*Afon Tafolog, Caerns	6·0	12·8	n.d.	8·5	121·2	0·35
*Mynydd y Dref, Conway, Caerns	4·8	35·5	n.d.	54·4	52·7	3·74
*Foel Fras, Caerns	4·6	16·1	n.d.	3·9	16·8	1·28
*Melynllyn, Caerns	4·3	78·7	n.d.	9·2	49·7	2·17
*Pen y Pass, Caerns	4·5	11·2	n.d.	13·2	6·9	0·47

† Data from Agrosto–Festucetum, in McVean and Ratcliffe (1962).
* Data from *Nardus stricta* habitats, in Chadwick (1960).

n.d, = no data

range of grasslands is their low pH, usually 3·5 to 6·0, and the low base status (particularly calcium) of the soils. As might be expected in a group of grasslands which extend from nearly sea level to 3000 ft (914 m) or more over the whole of the British Isles, the floristic composition is highly variable, and the delineation of communities difficult, especially at the association level. This is especially so in the upland acid grasslands where slight differences in topography and the presence of base-rich flushes can create differences in drainage conditions and edaphic factors which can result in several different grassland types occurring together within a small area. Consequently, there is little doubt that in these communities variation is continuous so that the classification of a particular grassland may some-times be a matter of opinion. For the purposes of this book, therefore, a conservative view is taken of acid grassland communities and the reader is referred to McVean and Ratcliffe (1962) and King and Nicholson (1964) for more detailed treatment of upland grasslands.

Acid grasslands are characterized by the following groups of species:

Grasses and sedges	*Festuca ovina, Agrostis tenuis, A. canina, A. stolonifera, Sieglingia decumbens, Nardus stricta, Molinia caerulea, Deschampsia flexuosa, Juncus squarrosus, Carex binervis, Luzula campestris* and *L. multiflora.*
Herbs	*Galium saxatile, Potentilla erecta, Campanula rotundifolia, Viola canina.*
Mosses	*Rhytidiadelphus squarrosus, Dicranum scoparium, Hylocomium splendens, Hypnum cupressiforme* and *Pleurozium schreberi.*

Unlike other grassland types, dwarf shrubs often grow mixed with herbaceous species to form the so-called 'grass heaths'. *Calluna vulgaris, Ulex europaeus, U. minor, U. gallii, Erica vagans* are the most common types of shrub in lowland heaths. In upland regions, *Calluna, Vaccinium myrtillus, V. vitis-idaea* and *Empetrum nigrum* are most frequent. In both the lowland and uplands, bracken (*Pteridium aquilinum*) is an important component of some grass heath communities. Gimingham (1972) describes in some detail the relationship between grass heaths and true ericaceous-dominant heaths, as well as providing a broad classification of European heaths which includes some of the grass heaths discussed in this book.

Classification

Acidic grasslands can be classified into six main types, four of which occur predominantly in the uplands, the other two mostly in the lowlands.

Upland *Festuca–Agrostis* grassland

This, the most widespread of all acidic grassland, occurs in Wales, the Pennines, the Southern Uplands and the Central Highlands and is related to the lowland *Festuca/Agrostis* grassland found on sandy heaths. *Festuca ovina* and *Agrostis tenuis* are the most abundant grasses, usually associated with *A. canina, Anthoxanthum odoratum, Galium saxatile, Potentilla erecta, Rhytidiadelphus squarrosus* and *Hylocomium splendens*. They are characteristic of free-draining brown earth forest soils of low base status or of leached skeletal soils. Humus is of the moder type and the pH range usually 4·7–5·2. In addition they are common in the foothills of the Southern Uplands between 100–1250 ft (30–381 m) and have been found up to 2700 ft (823 m) in the Highlands. This type is described by McVean and Ratcliffe (1962) as the species-poor facies of the Agrosto–Festucetum.

A species-rich variety of this main type is associated with similar brown earth forest soils of high base status, usually with a pH of 5·1–6·0. The humus is of the mull type and the soil has a good crumb structure. This type is recognized by the high frequency of *Festuca rubra, Plantago lanceolata, Thymus drucei* and *Trifolium repens*, together with less frequent but indicative species such as *Briza media, Koeleria cristata* and *Galium verum*. This variant of the *Festuca–Agrostis* grasslands is widespread in the areas mentioned previously, but never occurs extensively and is often associated with base-rich flushes or calcareous rocks. Many other species-rich variants of this type of grassland have been described by McVean and Ratcliffe (1962).

Nardus grassland

Nardus grasslands are developed on peaty podsols or on peat gleyed soils with from 10 to 30 cm of mor humus on the surface with a pH in the range 3·7–5·0. They occupy much of the upper slopes of the Cheviot Hills, and are widespread from about 1000–2300 ft (305–701 m) in northern England, Wales and the Highlands.

Nardus stricta and *Festuca ovina* agg. are the dominants and *Deschampsia flexuosa* is usually abundant. In addition to the dominants, these grasslands are characterized by the constancy of *Agrostis canina, A. tenuis, Anthoxanthum odoratum, Carex binervis, Juncus squarrosus, Galium saxatile, Potentilla erecta, Hylocomium splendens* and *Pleurozium schreberi*. This type has been described by McVean and Ratcliffe (1962) as a species-poor Nardetum sub-alpinum.

On leached podsolic soils and even on some brown earths of low base status, *Nardus stricta* may be co-dominant with *F. ovina* agg. and *Deschampsia flexuosa* in association with *Sieglingia decumbens*, *Carex binervis*, *C. caryophyllea*, *C. panicea*, *Luzula campestris* and *L. multiflora*. The range of this sub-type is usually pH 4·1–5·1.

Molinia grassland

Grasslands in which the dominant is the tussock-forming *Molinia caerulea* are found on peaty podsols or peaty gleys with up to 30 cm of peat. The soils are wet but never water-logged, lateral movement of water being essential for *Molinia*. Floristically they are similar to the *Nardus* grasslands, with *Festuca ovina* and *Deschampsia flexuosa* often co-dominant. They are well developed in all upland areas of Britain, commonly occupying the upper slopes and rounded tops of hills up to 1500 ft (457 m). They present a characteristically under-grazed appearance, the herbage being from 30–40 cm tall and often heavily tussocked.

Deschampsia cespitosa grassland

This type of grassland is found on poorly drained gleyed soils throughout upland Britain, especially where water stands in the winter, although it may not be very wet in the summer. There is frequently a thin layer of moder humus and the pH varies from 4·6–5·2. The tussocks formed by *Deschampsia cespitosa* are often so large that few other species are present, especially when the grassland is lightly grazed. Nevertheless, many of the species characteristic of the *Festuca–Agrostis* grasslands are found on the spaces between the tussocks and this type is only separated from the *Festuca–Agrostis* swards by the dominance of *Deschampsia cespitosa*. Similar grasslands are found on water-logged clays in lowland Britain.

Lowland *Festuca–Agrostis* dry heath grassland

Tansley (1953) used the term 'grass heath' to describe those grasslands on dry sandy soils in lowland England which occur on the same soils as *Calluna* heath. The studies of Farrow (1915) and Watt (1938) clearly show that the relationship of grass heath to lowland *Calluna* heath is essentially the same as that of the *Festuca–Agrostis* hill pasture to upland heath or heath moor. Grazing will convert heath into grassland, and grassland is invaded by dwarf shrubs in the absence of grazing.

Lowland grass heath is well developed in Breckland, on the commons of Surrey, Berkshire and Dorset and elsewhere in smaller quantity wherever dry, acid soils are present that have not been converted to arable land. Some of these soils, especially in the Breckland, have been covered with calcareous drift deposits which mask or alter the characteristic vegetation, and these grasslands are more correctly placed in the calcareous part of the classification.

Typically, these acid, lowland heaths are open communities, with tufts of *Festuca ovina*, *Deschampsia flexuosa* and *Agrostis tenuis* set in a matrix of lichens and mosses. Characteristic herbs in this community are *Polygala serpyllifolia*, *Thymus drucei*, *Rumex acetosella* and *Hieracium pilosella*.

The species list obtained from a variety of lowland heaths may be extensive, but an important character of acid grasslands is their floristic poverty, and any one grassland site is unlikely to support a great variety of species.

Agrostis setacea grassland

This type of grassland is restricted completely to south-west England and to a small region in South Wales. It occurs on dry heaths which have a thin layer of surface peat on a variety of geological formations, the soils being porous and relatively acid (pH 4·1–6·2) and podsolic in structure. Even though well developed on the serpentine rocks of the Lizard, the soil is an incipient podsol formed from a superficial deposit of silty fine sand.

Agrostis setacea grasslands are especially prominent following burning of heaths, usually occurring in communities which contain considerable quantities of ericaceous shrubs.

Molinia caerulea, *Sieglingia decumbens* and *Carex binervis* are common associates. In the unburnt heath, the grassland forms a mosaic with the dwarf shrubs, with a ground layer in which the mosses *Dicranella heteromalla* and *Campylopus flexuosus* are frequent.

Other Grasslands

Maritime and sub-maritime grasslands do not fit into the classification of grasslands described above and are best considered as a separate group They fall into three types:

Salt-marsh grassland

These develop from sand and mud flats on the coast by a process of accretion, the salt-marsh often exhibiting all stages in the sere from bare mud to brackish grassland. *Puccinellia maritima* is an important early colonist of the mud flats, but once accretion has raised the level of the mud surface, other species of the 'middle marsh' become prominent, for example *Limonium vulgare*, *Triglochin maritima* and *Halimione portulacoides*. Eventually the middle marsh is invaded by *Festuca rubra*, *Juncus gerardii* and *Plantago coronopus*, and as the period of tidal submergence becomes less, the familiar salt-marsh communities develop. They form valuable pastures for sheep.

The 'upper marsh' communities are characterized by the presence of *Festuca rubra*, *Agrostis stolonifera*, *Carex flacca*, *C. panicea*, *C. nigra*, *C. extensa* and *C. distans*. In the southern half of Britain, *Spartina × townsendii* often replaces *Puccinellia maritima* as a salt-marsh pioneer species, and in many places this grass forms luxuriant swards which are cut for hay or grazed by cattle.

Dune grassland

Sand-dune systems are sub-maritime communities, in the sense that they are not subjected to periodic inundation by the sea and are not strongly saline. Many seral stages are usually represented on any sand-dune system and although some of the pioneer stages are completely dependent on grass species, these are not used as grasslands in the agricultural sense.

Grasslands in the dune system which are grazed by sheep and cattle are those developed on the 'dune-meadow' and 'dune-slack', the former occurring on the landward side of the dune system with a water table well below the surface, the latter with a water table which is near the surface for at least part of the year.

In the dune-meadow a vegetation similar in many respects to that of calcareous grassland develops, with *Festuca ovina*, *F. rubra*, and *Agrostis tenuis* dominant, and *Thymus drucei*, *Hieracium pilosella*, *Galium verum*, *Sedum anglicum*, *Cerastium holosteoides*, *Veronica chamaedrys*, *Viola tricolor* ssp. *curtisii*, *Trifolium repens* and *Ononis repens* forming a grassy sward. Mosses are often abundant, especially *Tortula ruraliformis*.

Dune-slack vegetation varies according to whether it is developed on a non-calcareous or calcareous dune system. In the former it is composed of a range of communities which show a similarity to certain types of acidiphilous valley mire communities, with *Carex nigra, C. echinata, C. curta, C. rostrata, Agrostis stolonifera, Eleocharis palustris, Hydrocotyle vulgaris* and *Ranunculus flammula*. In calcareous dune-slacks there is a mixed sedge growth with *C. nigra, C. flacca, C. panicea, C. pulicaris, C. demissa* and many dicotyledons, including *Dactylorchis incarnata* and *Epipactis palustris*.

The reader is advised to consult Ranwell (1972b) for a detailed account of salt-marsh and sand-dune vegetation.

Cliff-top grassland

This type of grassland which develops on cliff-tops is partly determined by the nature of the underlying rock formation and partly by the intensity and duration of salt-spray drift. There is some disagreement among ecologists as to whether cliff-top vegetation represents an edaphic or climatic climax or whether, like most other grasslands in Britain, it is a biotic plagioclimax.

On exposed rocky outcrops near the sea, a light, closely grazed turf develops with *Festuca rubra* abundant, with *Plantago maritima, P. coronopus, P. lanceolata* and *Armeria maritima* frequent. On less exposed cliffs, these species remain constant, and they are often accompanied by *Silene maritima, Sagina maritima, Sedum anglicum* and *Cochlearia danica*. In western districts of Britain, *Festuca rubra* grasslands are widespread and *Scilla verna, S. autumnalis* and *Erodium maritimum* are often well-established constituents.

On calcareous cliffs, the range of species differs little from that found in inland grasslands on similar soils, except that the maritime element in the flora includes many species which are rare away from the coast. This is especially true of the cliff-edge grasslands on the serpentine rocks of the Lizard peninsula, Cornwall, where Mediterranean and southern European species unique in Britain flourish, e.g. *Trifolium strictum, T. incarnatum* ssp. *molinerii* and *T. bocconei*. The reader should consult Malloch (1971) for a detailed account.

3 Grassland and scrub as ecological systems

The purpose of this chapter is to discuss grassland and scrub within the context of current ecological thinking. The complexities of ecological science are such that there are many difficulties to such a discussion, and these difficulties are increased by the fact that there is no theory of ecology which is either sufficiently rigorous or universally accepted. There are considerable differences of opinion among ecologists on most theoretical aspects of their science, while some fields, such as population dynamics, have been notable for the polemical writings of different schools of theorists. Nevertheless, the ecology of grassland can only be viewed in perspective if it is considered within a theoretical framework. Studies on grassland have already added much of importance to the theory of ecology and doubtless have much to contribute in the future.

The term 'ecosystem' implies a dynamic and functional approach to ecology. Such an approach is essential and appropriate, for ecology is a dynamic science. However, it may be helpful to discuss some static and classificatory concepts of ecology as they relate to grassland, before considering its dynamics as a functional system. Populations of both plants and animals are usually found to be grouped together, or mixed, in space and time; such assemblages are frequently called *communities*. Communities may be large or small, but the biosphere (all the plants and animals on earth) is conceived as comprising a small number of basic communities called *biome-types*, of which the vegetation component is often called a formation-type or type of vegetation (Tansley, 1946). Whereas some ecologists conceive formation-types as being developed entirely in relation to climate (e.g. Whittaker, 1970), others recognize edaphic and other kinds of formation-type as well (e.g. Tansley, 1953). Each biome-type is physiognomically distinct, although the species of which it is composed,

70

including the dominant species, may not be the same in different parts of its range. The communities of a biome-type dominated by different species are the *biomes* which together comprise the different biome-types. A useful short account of biome-types is given by Whittaker (1970).

The concept of *dominance* relates to the community structure of vegetation in particular; it is perhaps a less useful idea when considering animals although Odum (1959), for instance, applied the concept to any 'important' species population in a community. It is usual to classify as dominant those species in a community which most influence the other species and are least influenced by them. Although dominance is easily appreciated by field naturalists and ecologists it is a difficult idea for theoreticians to quantify. However, McNaughton and Wolf (1970) have been able to relate dominance to the concept of *niche* and to species-richness of communities. Niche has been so variously used in ecological writings that its present usefulness, both as a descriptive term and as an idea, is somewhat doubtful. Originally coined by Grinell (1924) and used in a largely geographical and spatial context, the word niche was used independently by Elton (1927) in an attempt to describe the *functional* position of a species population in a community, particularly in relation to its feeding habits. Hutchinson (1957) conceived of niche as a hypervolume formed from values of all relevant ecological parameters, thus developing Elton's idea. Recent developments of niche theory, in attempts to produce greater precision, have merely tended to irrelevance and incomprehensibility (e.g. Vandermeer, 1972). Niche is related to the much simpler concept of *habitat*, which is the place where a species lives. The word is often used indiscriminately in this sense and as a synonym for *biotope*, which is an area of land or water with defined characteristics (e.g. Macan, 1963). In some cases 'niche' appears to be used where 'habitat' is appropriate (e.g. Pielou, 1972).

In a world-wide sense grassland forms a single biome-type, although montane grasslands are usually regarded as forming part of an arctic–alpine biome-type. Classification of communities into biome-types is based on physiognomy and largely determined by climate (Fig. 3.1). The classification does not imply any dynamic relationship between different biome-types, but such a relationship does exist and will be discussed in a later part of this chapter. The grassland biome-type consists of a number of biomes, which can be defined in terms of physiognomy and environment. In Britain grasslands have developed under a temperate climate of Atlantic type, but are generally regarded as being essentially anthropogenic. In other parts of the world this is not necessarily so. For instance, the Steppes and Prairies of continental Eurasia and central North America respectively

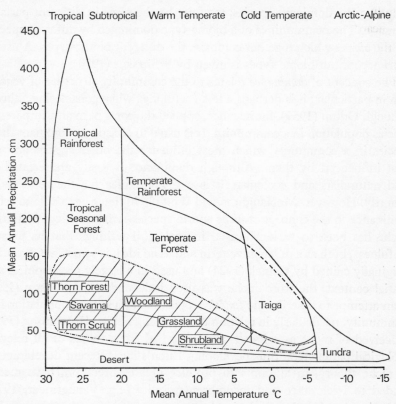

Fig. 3.1 Relationships of world biome types to temperature and precipitation. (*Courtesy of* Professor R. Whittaker.)

are thought to be covered by climax vegetation, that is, the type of vegetation which would develop naturally under the prevailing climatic conditions. These grasslands are also correlated with the occurrence of chernozem soils and the distribution of deposits of loess from which the soils were formed (Bridges, 1970). All grassland biomes are dominated by grasses and grass-like plants, such as sedges, and the most frequent life-form is the hemicryptophyte (Raunkiaer, 1934) (Table 7.1). Grassland differs markedly in its physiognomy from other biomes, such as forests, dominated by various kinds of phanerophyte, and deserts, variously dominated by therophytes (annuals), succulents and geophytes.

The floristic characteristics of grassland are discussed in Chapter 4 and those of scrub, the shrubland biome, in Chapter 6. The classification used is based primarily on criteria of adaptation and dominance and makes relatively little use of phytosociological methods; Shimwell (1971c) has recently reviewed the history and methodology of the description and classification of vegetation. The dominance of hemicryptophytes in grass-lands results in their having a relatively simple vertical structure. Thus grassland contains but two 'habitat types' (Elton and Miller, 1954; Elton, 1966) – ground zone and field layer – although the latter may not always be present on every grassland area. Besides the ubiquitous 'air above', scrub and woodland include one and two additional 'habitat types', low canopy (of trees), and low and high canopy, respectively. This classification, which is really one of biotopes, is necessarily crude but probably reflects fairly faithfully the importance of the different 'habitat types' for animals, al-though this has not been demonstrated.

Although it is convenient to recognize biomes and other communities the biosphere is a continuum. The nature of the boundaries between one community and another is a feature of considerable theoretical and practical importance. At one extreme very gradual changes between com-munities can be recognized with transition zones which are broad in extent; these are known as *ecoclines*. Ecoclines are of interest and import-ance (Whittaker, 1970) but are not readily observable in the British Isles, partly because the areas of transition are often so large in the case of major ecoclines, and also because they occur in natural areas where the influence of man has not been great. In regions such as Britain the natural vegetation has been much changed and gradual transitions between one community and another have been fragmented or destroyed. Abrupt transitions between communities, on the other hand, are more frequent in anthro-pogenic landscapes than natural ones because modern agriculture and forestry, for example, tend to form sharp boundaries between adjacent communities. Such abrupt areas of transition are called *ecotones*, but the term is also frequently used to denote quite gradual zones of transition, although these are assumed to be less wide than are the adjoining two communities separated by the transition zone (e.g. Odum, 1959, p. 278). Boundaries between grassland and scrub, or between grassland or scrub and woodland, are examples of ecotones. Their most important ecological characteristic is the so-called 'edge effect'. Many ecotones are richer and more diverse in species of both plants and animals than either of the communities they separate. This is not only because they may contain species from both communities but also because they may have species of their own, that is, characteristic of the ecotone rather than of the grassland, scrub or woodland communities themselves. This is not surprising, because it is likely that the clearings and open spaces associated with large areas of climax woodland were the natural environment of many grassland plants and animals. Such woodland was probably uneven-aged with open communities forming in gaps left by dying and falling trees. The importance of ecotones is recognized in the habitat-type classification of Elton and Miller (1954). The edge-effect is often particularly pronounced in the case of birds. However, it is by no means a universal phenomenon; cases are known where an ecotone is less rich in species than the communities it divides.

Mountain grasslands differ from lowland grasslands in Britain in that they are maintained, at least to a considerable extent, by climate rather than by man and his grazing animals. Some coastal grasslands, especially those on the tops of cliffs, are also maintained in this way. These grasslands thus partly resemble those of the Steppes and Prairies and some African grasslands which have developed in response to climates too severe for forest to develop. In Britain it is widely supposed that the natural vegetation of the lowlands since the last (Weichselian or Würm) glaciation has been deciduous broadleaved forest (Chapter 1). Such forest is regarded as having been the *climatic climax* vegetation of our region during the Flandrian period, in which the three phases of recolonization, dominance and clearance of such species as oak (*Quercus* spp.) are usually recognized. Although there is some controversy about the exact extent and exclusiveness of the forest it is known that the phase of clearance of the woodland took place from Neolithic times onwards (Godwin, 1956).

Succession

The significance of succession is basic to understanding the grassland and shrubland ecosystems. Much of the earliest work on this subject was done by Clements (1916), who had a considerable influence on the thinking of later ecologists. Succession is the changing of communities in time as they respond to conditioning of their environment by their own activities (*autogenic succession*) or through the action of some external agency, such as changing climate (*allogenic succession*). The sequential communities which form the succession are collectively known as a *sere* and individually as *seral stages*. A complete sere, from the first colonization of new land to the development of climax vegetation, may be termed a *prisere* and the type of succession is primary. Succession frequently begins, however, with the destruction of an earlier community, for example the felling of a woodland. The complete sere in such a succession may be called a *subsere* and the succession itself is secondary. It may be supposed that a short-lived seral stage in the succession to forest climax was a community having some of the characteristics of grassland. After woodland has been felled a community dominated by grasses and tall herbs may develop but in the absence of restraining factors it rapidly changes back to a form of woodland. The phenomenon is well known in tropical forests where clearance is made for temporary agriculture; reversion to secondary forest is rapid. In much of the forest cleared in Britain in neolithic and historic times a restraining factor in the form of man and his grazing animals prevented reversion to forest. In ecological terms the tendency to form climax forest has been deflected. Grassland is neither a seral stage, nor a climatic climax, it is a deflected climax or *plagioclimax*. The sere culminating in a plagioclimax is called a *plagiosere*. Many of the problems of managing lowland grasslands result from their being plagioclimaxes.

Most scrub, or shrubland, on the other hand, is not a plagioclimax but a seral stage. The development of scrub from grassland may best be regarded as the start of a subsere in which cessation of management, in many cases by the removal of grazing animals, has initiated secondary succession. Development of scrub from cleared woodland may be thought of in a very similar way. This ecological difference between most grassland and most scrub is fundamental to problems of management and necessitates a different approach to the conservation of these two very different types of community. However, it is by no means impossible for scrub communities to be maintained dynamically as plagioclimaxes just as grasslands are maintained in dynamic equilibrium between two opposing and normally equal forces – successional changes towards scrub (and ultimately woodland) and management, usually the pressure of grazing animals.

Not all scrub, however, is a seral stage. Maritime scrub is frequently a climax community, just as are some maritime grasslands. There has in the past been some controversy as to the nature of climax vegetation, particularly over the question whether a single climax (*monoclimax*) or several climax communities (*polyclimax*) can occur in the same general region under the same climate. Shimwell (1971c, p. 128 seq.) has recently discussed this in detail and shown that the argument is essentially semantic. Climax vegetation is not everywhere invariably the same and its nomenclature is relatively unimportant. The phenomenon of *convergence*, however, has been frequently observed in different seres which tend to develop towards the same climax. In many seres the later seral stages are dominated by fewer species than the earlier ones (Fig. 6.1), so that convergence of dominants, at least, takes place. Another ecological process which may occur in the formation of some shrubland communities is the phenomenon of *retrogression*. It is believed that hazel (*Corylus avellana*) scrub in

Derbyshire has been so formed by the removal of trees from climax woodland (Shimwell, 1968).

An obvious characteristic of autogenic succession is that nutrients tend to accumulate in the soil with *eutrophication* increasing as the sere advances towards climax. Green (1972) has argued that in the management of seral stages, such as grassland, nutrients should be removed from the system to counterbalance eutrophication. Grazing, mowing and burning have this effect in many cases, whether it is realized or not. Unfortunately, Green only considers the vegetation, it being true, of course, that relatively much less information exists about the fauna. There is, however, good evidence that taller grasslands support more animal species than short ones and the floristic diversity of such grasslands may be quite unrelated to their biotic diversity. This is discussed later in this chapter and also in Chapter 7. Much more work needs to be done on these problems before a satisfactory theoretical, as well as practical, basis for management is achieved. In particular, great attention needs to be given to questions of scale and duration, especially in relation to *time*.

Grassland Energetics

In grassland, as in all other communities, the radiation energy of the sun is transformed by photosynthetic plants, the grasses, sedges, dicotyledonous herbs and mosses, into matter and may subsequently be transformed by animals feeding on the plants and on each other. In the classic concept of *trophic levels* formulated by Lindeman (1942) levels of energy transfer are conceived to be present in the ecosystem (Fig. 3.2). Solar energy passing through the atmosphere and entering the ecosystem λ_0 is transformed at

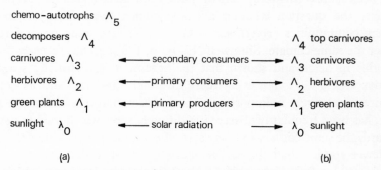

Fig. 3.2 Trophic organization of ecosystems. (Lindeman, 1942.)
(a) whole ecosystem
(b) ecosystem excluding decomposers and autotrophic organisms

the primary producer level (Λ_1) into photosynthetic plant tissue, with loss of energy in accordance with the second law of thermodynamics. Animals feeding on plants transfer energy in a similar way. It is sometimes convenient to separate the decomposers and chemo-autotrophic bacteria (trophic levels Λ_4 and Λ_5) from the schematic representation (Fig. 3.2a) and to include a trophic level of 'top carnivores' as Λ_4 (Fig. 3.2b). Trophic organization is conceptual rather than descriptive, because any particular population of animals may be partly phytophagous and partly predacious. A more descriptive representation of the organization of an ecosystem is the *food-web*, often expressed diagrammatically. In a food-web diagram the relationships between plants and prey and predators are shown as a network of lines joining the different organisms. It is supposed that communities with many links in their food-webs are more stable than those with few. Food-webs are an elaboration of the earlier idea of food-chains; both are discussed in most ecological text-books (e.g. Elton, 1927; Allee *et al.* 1949; Odum, 1959). When the trophic organization of any ecosystem is examined it is evident that the relationships between levels can be expressed in the form of a *pyramid*. Ecological pyramids may be of numbers of individual organisms, biomass or fixed energy (Fig. 3.3). The last is often the most meaningful in studies of energetics.

The management of grassland, in most cases, represents a loss of matter, and therefore energy, to the system. Under conditions of dynamic equilibrium the energy removed from the grassland as hay (in meadows) or

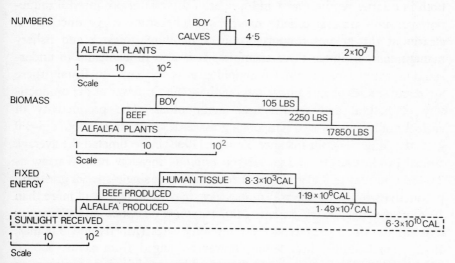

Fig. 3.3 Ecological pyramids. Each pyramid represents the relationship between trophic levels in the ecosystem. (*Courtesy of* W. B. Saunders Co.)

sheep and cattle (in pasture) is balanced by the energy put into the system in the form of solar radiation. In the absence of management matter accumulates in the system. Grasses readily form tillers, or lateral shoots, and it is these which become compact grass turf under intensive management, especially grazing. In the absence of management tussocks are often produced, particularly by certain grasses, such as species of *Dactylis*, *Molinia* and *Deschampsia*. Under these circumstances the nature of the available plant material in the system changes from predominantly fresh, living foliage to predominantly old, and especially dead, material. The animal community changes from one consisting mainly of primary consumers to one in which saprophagous, decomposer species predominate. Even so, the animals are unable to decompose all the plant litter and dead plant material accumulates as succession proceeds. The two types of animal community both possess predatory and parasitic species and, of course, grade into each other. Both are important, not only in a functional sense, but in what they contribute to the interest and richness of the grassland ecosystem in the broad sense.

Production

The fixing of energy at any trophic level in the ecosystem is called production. It is usual to distinguish between primary production (of photosynthetic plants) and secondary production (of animals) and to measure both in calories per unit area and weight of dry matter produced. Production per unit time is called *productivity*. The study of productivity is clearly of the greatest importance in agriculture, forestry and fishery management, to quote three examples. It is also fundamental to understanding many aspects of the functioning of ecosystems. In Britain there are many studies of agricultural productivity but far fewer of the productivity of natural and semi-natural communities. The productivity of agriculturally unimproved temperate grasslands ranges from about 0·5–3·0 g/m²/day (Fig. 3.4); for instance Traczyk (1968) has estimated the average annual productivity of a *Deschampsia cespitosa* meadow near Warsaw as 2·0–2·1 g/m²/day (*c.* 750 g/m²/yr). A recent study of single-species grassland productivity in the northern Pennines gave maximum figures of more than 10 g/m²/day for *Lolium perenne*, *Anthoxanthum odoratum* and *Cynosurus cristatus* for short periods in May on some sites (Morris and Thomas, 1972). Total annual production, however, ranged from about 500 to 700 g/m². The range of figures for grassland productivity is much less than is achieved in more specialized ecosystems, such as intensive tropical

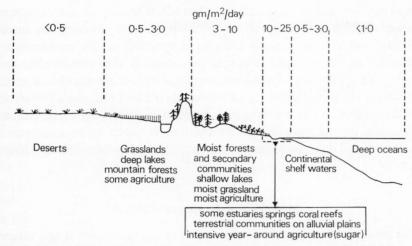

Fig. 3.4 Productivity of some ecosystems. (*Courtesy of* W. B. Saunders Co.)

agriculture. Primary productivity is apparently very inefficient, as a maximum of about only 1% of solar radiation energy is fixed by green plants, the remainder being lost to the ecosystem. The efficiency of energy transfer increases at the higher trophic levels and some insect parasitoids are particularly efficient in this respect.

Maximizing productivity is a major aim of agricultural science, and indeed of much human endeavour. In farming, forestry and fishery management this means channelling energy into forms which can be exploited by man; an example in freshwater is given by Odum (1959; frontispiece). Agriculture and kindred activities require transfers of energy to be in a few, broad channels, whereas nature conservation involves transfers of smaller amounts of energy through many, narrow pathways. In these simple ecological factors lies the incompatibility of wildlife conservation and modern intensive agriculture.

Diversity

The many, narrow pathways of energy transfer in a natural ecosystem are one aspect of the *diversity* of the system. Species-diversity of grasslands, as of other ecosystems, is a measurable quantity, high values of which Margalef (1963, 1968) and other ecologists regard as being correlated with stability, maturity, resistance to change and low productivity in the ecosystem. Grasslands and shrublands are both very rich in animal and plant species, but forests, especially tropical rain-forest, are generally accepted as being the richest ecosystems. A distinction should be drawn between *richness*, which is the number of species present, or breeding, in an ecosystem and *diversity*, which includes a measure of the distribution of numbers among the species in that ecosystem. A very considerable literature on the diversity of different communities and associations of species in various biotopes has been published in the last few years but its immediate relevance to conservation problems is not always clear. In many ways the maintenance of species richness is much more important in conservation than increasing diversity. This is discussed by Morris (1971a).

Models and Systems Analysis

The concept of the trophic organization of the ecosystem is one attempt to simplify an extremely complex series of interrelationships in order to study the functioning of the system as an entity. In any ecosystem some of the species-populations are of much greater importance to the dynamics of the system than others. Thus the dominant grass species and grazing animals are of primary importance in the dynamics of grassland. It is often possible to simplify ecosystems by concentrating on the components which are important to its functioning and one method of doing this is by mathematical modelling. Models, mathematical or of other kinds, 'are devices to assist thinking' (Solomon, 1971) and 'are conceived as expressions of thought and ideas in language, with all the charms and the dangers that ideas may have' (Reddingius, 1971). Models of ecosystems can be formulated and tested against the actual system, with the possibilities of prediction and control in mind. Mathematical ecologists are at pains to emphasize, as does Jeffers (1972), that mathematics, statistics and computers are merely tools for ecologists to use and exploit. However many ecologists would not agree that the *most* important impact of mathematics arises from the formulation and use of models. Mathematics is at least as important in the analysis of field data and experimental results.

The approach to ecology through systems and systems analysis is relatively recent, but has considerable potential. In this context 'system' does not mean the same thing as 'ecosystem', although both terms have a functional connotation. The application of systems analysis to ecology has been discussed by Watt (1966) and his co-workers.

Population Dynamics

As we tend to regard grasslands as being maintained dynamically in a steady state it is possible to lose sight of the fact that these ecosystems are composed of species-populations which are liable to marked fluctuations in numbers, as well as being subject to control. This is particularly so with animals, although the controlling factor of man in determining the populations of large herbivores makes us less aware of the fact. Recent advances in the theory of population control have stressed that the same factors do not act on different populations. Much of the controversy associated with such names as Nicholson, Thomson, Andrewartha and Birch, Solomon, Varley and Milne has been resolved by the realization, from many detailed studies, that weather, competition, parasites and predators, foodplant limitation, behaviour and social feedback may all be controlling factors in particular cases.

Understanding the population dynamics of grassland animals is essential for the conservation of individual species. It also provides an extremely important method of examining the nature and composition of communities in a functional sense. The study of one species inevitably involves the examination of its prey, parasites, predators and competitors and so the whole community. Unfortunately, more general methods have to be used in addition, because of the impossibility of making detailed studies of the population dynamics of more than a few species.

4 The grassland ecosystem: botanical characteristics and conservation interest

As we have seen in the previous chapter, grasslands in temperate regions can only remain as such if the annual production of vegetative material is removed either by grazing or by cutting. It is important to remember this dynamic aspect and the part which management plays in determining the floristic composition at any given point in time. How various management techniques can be used to alter the botanical composition of grasslands will be dealt with in later chapters.

Factors Affecting Botanical Characteristics

The main factors affecting the distribution of grassland plants throughout Britain are soil, water table, geographical location and management. These factors are often very much interdependent and frequently difficult to analyse separately. Field work on assessing the relative importance of each is mainly confined to the work of the agriculturist attempting to gain maximum production from a relatively few species which are of economic importance.

Soil

Grassland occurs on a wide range of soil types, from those derived from the basic geological formations of chalk and limestone – the latter merely a harder and impure form of chalk – to those on rocks which weather very slowly and produce little soil, often of low fertility. The distribution of chalk and limestone in Britain is illustrated in Fig. 2.1, which is based on the solid geology. This does not take into account the fact that in parts of Britain, and in particular lowland Scotland, calcareous rocks may be masked by secondary material deposited during the glacial periods. The chalk and limestone grasslands *per se* and much of the neutral grasslands are largely confined, therefore, to areas in southern Britain.

In 1865 Nägeli commented on the relationship of competition between closely related species and their soil preferences. He instanced the case of *Achillea atrata* and *A. moschata*, the former occurring on limestone and therefore termed a calcicole, whilst the latter is widely distributed on siliceous soils and is a calcifuge. Tansley (1917) studied a similar pair of congeners, *Galium saxatile* (calcifuge) and *G. pumilum* (calcicole), sown together on various soil types, and found that both species would germinate on all soils but that the seedlings of *G. saxatile* suffered from chlorosis on calcareous soils and were rapidly killed off through competition by *G. pumilum*. Under acidic soil conditions the reverse occurred but not with as marked an effect. Both species could survive on the 'wrong' soil but only if competition from the other species was excluded. Salisbury (1920) suggested that the many factors involved could be grouped under two headings – chemical and physical. The latter are determined mainly by the water regime and he quoted Walker (1920), who showed that the same features which make for dryness also ensured good aeration, and this tended to reduce acidity. Salisbury attributed the chemical effects to the calcareous soil being more favourable for the growth of suitable bacteria, which are necessary for the prevention of toxic effects brought about by decay, and also for the production of nitrates, essential components of plant metabolism.

That the pH of the soil is not directly responsible for the calcicole-calcifuge concept was demonstrated by Grubb *et al.* (1969) who worked on a chalk heath community where both calcicole and calcifuge species grow together. They showed that both groups of plants had their roots growing healthily in mildly acidic soils of pH 5 to 6. They suggested that 'competition, combined with physical factors such as poor aeration and too low temperatures, are most important in excluding strict calcicoles from most soils of pH 5 to 6 in Britain'.

The interaction of aluminium with calcium in determining pH is documented by Hutchinson (1945) and is given by Rorison (1960) as the reason for the failure of *Scabiosa columbaria* and *Asperula cynanchica* seedlings to establish on acid soil.

Calcium is the major active element present in soils and it is not surprising that the early writers were concerned with its effect on plant distribution and the relationship between this element and acidity measured as pH. Nutritionally, calcium is rarely deficient, but because of the cationic relationship with pH it has a marked effect on the availability of elements such as aluminium and manganese. These cations are more readily available under acid conditions. Clymo (1962) suggests that this may be a factor accounting for ecological differences between *Carex demissa* and *C. lepidocarpa*. Both sedges tolerate a pH range of 4·5 to 7·0 but *C. demissa* is usually found where the calcium content is less than 30 p.p.m. although this species can withstand higher levels of aluminium.

Various workers (Bradshaw *et al.*, 1958; Jefferies and Willis, 1964) have noted the relationship between certain plant species, calcium concentration and their ecological distribution. Snaydon and Bradshaw (1969) have further shown that, even within a species (e.g. *Trifolium repens*), there are natural populations which differ in response to calcium, magnesium and potassium. These differences are related to the soil type from which the populations were chosen.

Under conditions where agriculture is not working towards greater productivity, nitrogen appears to exert an influence only on those grasslands on 'rich' (high base status) soils. Tables 2.1, 2.3 and 2.4 show that calcareous and neutral soils have in general a higher base status, particularly of phosphorus and potassium, than acidic soils. These base-rich soils, many of which have been lost to agriculture, are typified by the presence, as a sub-dominant, of *Lolium perenne*, a species absent from poor soils. The classification of grass species according to soil preference was carried out by the agriculturist many years ago but, because herbs are generally regarded as 'weeds', no comparable classification which includes them has yet been made.

Water

All plant growth is dependent upon a supply of water and, because of its overriding effect on productivity, agriculturists have made careful studies of soil moisture and use the term 'field capacity' as a level from which changes can be calculated. Field capacity is defined as 'the moisture content of the soil when it is holding all the water it can against the force of gravity'. For sandy soils the concept of field capacity is not as precise as for heavier soils, because, after the initial rapid drainage, a very slow drainage may persist over long periods (Anon., 1954). For further information on the theoretical and practical aspects of soil moisture in relation to grassland productivity, the reader is referred to Penman (1948, 1952), and Schofield and Penman (1949).

Ecologically it is not the total reserve of water in the soil which is important but only that quantity available to the plant. On the other hand if the water content is too high, as can occur on ill-drained soils, oxygen supply may be a limiting factor. The two major factors affecting soil moisture are rainfall and the physical nature of the soil. Wilcox and Spilsbury (1941) have demonstrated that the wilting coefficient increases with the increase in the clay fraction of the soil, and Williams (1968) concluded that the water content and suction force, a measure of soil water potential, were related to the humus colloids (Fig. 4.1). The degree and extent of fluctuations in soil moisture may be as important as the gross differences between soils with varying water tables in determining floristic composition.

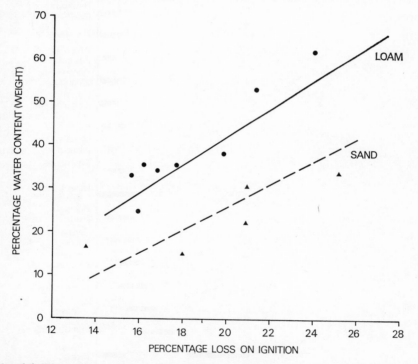

Fig. 4.1 The water content at 5 atm and the percentage loss on ignition. (*From* Williams, 1968.)

Whilst certain herbs may have deep-rooting systems which can exploit the water-holding capacity of lower depths, and thus be cushioned against surface fluctuations, the grasses are shallow-rooting and therefore measurements of soil moisture may be meaningless if they encompass too great a depth. Grasslands on the chalk downs of southern England are considered to be relatively xerophytic for although there is sufficient available water for much of the year, the critical drought period occurs at the time of maximum growth and only those species adapted to these conditions can flourish. Anderson (1927) has shown that for many chalk grassland species the greatest root development is in the upper 23 cm (Fig. 4.2). Most of the root growth can be directly related to the moisture content of the soil; fewest roots occur in the driest zone (30–38 cm) while deep-rooting species

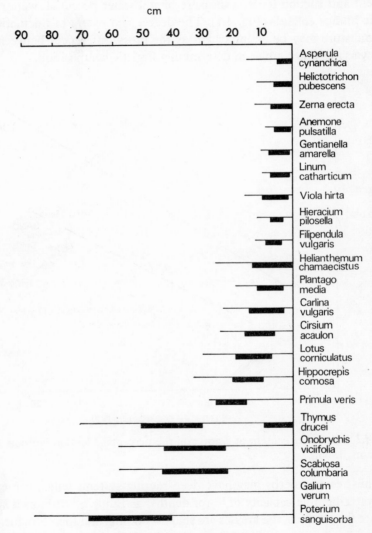

Fig. 4.2 Diagram of rooting depth of certain chalk plants showing area of maximum development of feeding roots (thick line) and maximum depth of penetration (thin line). (*After* Anderson, 1927.)

have maximum root development below the 40 cm zone. However, in the top 8 cm large fluctuations occurred with very low minimal values, some below the point at which wilting was observed under laboratory conditions. That wilting was not observed in the field was attributed to the slow capillary action in chalk soils first demonstrated by Ansted (1865).

Tansley's view that water availability was the major factor affecting the flora of chalk grassland has been challenged by Bunting and Elston (1966), who combined a study of soil water availability with applications of nitrogen and potassic fertilizers to chalk grassland. They concluded that nitrogen, and not water, was the major limiting factor. The role of nutrients has been extensively studied by the agriculturist whose object is high productivity. The ecological implications of the mineral nutrition of plants has received relatively little attention. Much of the work which has been done was reported at a British Ecological Society Symposium (Rorison, 1968).

Geographical location

Various theories have been put forward as to the origin and the geographical affinities of the British Flora and accounts of some of these are to be found in Forbes (1846), Watson (1847), Reid (1911), Matthews (1937) and Pigott and Walters (1954). Matthews groups some 700 species into 12 geographical elements; these are shown in Table 4.1 with the number of

Table 4.1 Total number of species and number of grassland species in the 12 geographical elements (*after* Matthews, 1937)

Geographical element	Total no. of species listed by Matthews	No. of species associated with grassland
Mediterranean	38	0
Oceanic Southern	82	14
Oceanic West European	87	15
Continental Southern	130	40
Continental	88	32
Continental Northern	97	32
Northern – Montane	31	3
Oceanic Northern	23	2
North American	6	1
Arctic – Subarctic	28	0
Arctic – Alpine	75	1
Alpine	10	0
Total	695	140

grassland species in each category. The three major elements are Continental Southern, Continental, and Continental Northern. This is mainly a reflection of the association of many grassland species with either the chalk of southern England or the Carboniferous limestone of northern England. The northern-montane and arctic-alpine elements are represented by species such as *Primula farinosa*, *Trollius europaeus* and *Bartsia alpina*, which reach their lower altitudinal range in a few enclosed grasslands of northern England.

Altitude effects are mainly determined by temperature. However, the importance of rainfall should not be overlooked because with increased precipitation montane plant associations may descend into lower altitudes, cf. *Bartsia alpina* in northern England.

Management

This is often the major factor affecting the floristic composition of lowland grasslands. Of the 17 997 970 a. (7 289 178 ha) of grasslands (recorded in the Annual Returns for 1971) in England and Wales, only a very small proportion is of high conservation value for wildlife. The 3 545 000 a. (1 435 725 ha) of clover and rotation grasses (i.e. grasslands under 7 years old) as well as much of the 9 789 000 a. (3 964 545 ha) of permanent pasture, (i.e. over 7 years old) are dominated by *Lolium perenne*. Indeed, of the 160 or so species of grasses found wild in Britain, only four or five are utilized to any extent by the farmer, e.g. *Lolium perenne*, *Dactylis glomerata*, *Phleum pratense* and *Festuca pratensis*. These, together with *Trifolium pratense* and *T. repens*, are the most common constituents of the grass seed mixtures sown at the present time. The number of plant species sown by the farmer has progressively declined with advances in agricultural technology. The earliest re-seeding of land down to grass was achieved by using the sweepings from the hay barn, which would therefore include many of the indigenous grassland species. This, together with the large tracts of unimproved grasslands on the chalk, limestones, and in river meadows, probably resulted in little, if any, decline in species abundance before 1600. In the seventeenth century clover seed was imported into Britain from Holland and by 1700 sowing of clover, either alone or with grass seeds, appears to have been widespread (Fussell, 1964). After 1700 and especially from 1900 onwards, the numbers of many attractive grassland plants such as the green-winged orchid (*Orchis morio*) and autumn crocus (*Colchicum autumnale*) fell rapidly as the thin soils of the chalk and the previously ill-drained meadows were brought into arable cultivation. When these arable areas are returned to grassland, simple seed mixtures, specifically grown for the purpose, are nearly always used.

The ecological effects of grazing and cutting can be considerable, particularly if either is continuously used as the major process over a period of time. The chalk downs are well-known examples of grazed grassland resulting in a short, compact sward. The alpine hay meadows are equally well known, but in Britain our traditional hay meadows are more scattered and certainly less numerous. They occur in the dales of northern England and also in the south, but only where this form of management has been prescribed by law. An example is Lammas Land where grassland similar to continental meadows can sometimes be found. It is only in such areas, which have been cut for hay for several hundreds of years, that one can find such continental meadow species as *Ranunculus auricomus* and *Anemone nemorosa*. In Britain these species are sometimes thought to indicate recently cleared woodland but this is almost certainly erroneous in the case of grasslands utilized as a hay crop. They quickly die out if a change of management from hay cutting to grazing is carried out. Because there are so few traditional hay meadows in Britain *Arrhenatherum elatius* is not an abundant meadow species, as on the Continent, but is common on roadside verges. It is well known that this species is intolerant of grazing and this was well demonstrated by Baker (1937) in his classic comparison of Port Meadow with Pixey and Yarnton Meads, the former having been grazed and the latter cut for hay, probably since 1085. Baker found *Arrhenatherum elatius*, as well as *Bromus commutatus* and *Sanguisorba officinalis*, abundant at Pixey and Yarnton but absent from Port Meadow. A similar study of two Worcestershire meadows, one cut, the other grazed, but with a recorded history going back for only 60 years, showed a greater number of *Carex caryophyllea* per unit area on the grazed meadow. The effect of changing from hay to spring/summer grazing is very dramatic in reducing the numbers of *Fritillaria meleagris* – a plant closely associated with the hay meadows of the upper Thames basin. These examples illustrate how grazing assists those plants which, like *C. caryophyllea*, can spread readily by vegetative means but is disadvantageous to those species, like *B. commutatus* and *F. meleagris*, which are dependent on seed production. However very little seed will be set where there is a regime of continuous grazing.

Both *Arrhenatherum* and *Trisetum* associations are regarded by Passarge (1969) as characteristic of fertile soils, whilst *Festuca/Agrostis* associations are typically found on poor soils. In Britain, much of the grassland on the former has been ploughed and utilized for arable production, particularly in times of war, leaving only the thinner and relatively poorer soils under continuous grassland cover.

Until recently the objective of grassland management was optimum productivity; that is the land was encouraged to produce as much grass as its natural potential allowed. For example, grazing on common land was often controlled to avoid overstocking and farmyard manure was applied to hay fields to compensate for the loss of nutrients removed in the crop. The advent of manufactured inorganic fertilizers and herbicides coupled with new and improved agricultural machinery over the last 100 years has allowed maximum, as opposed to optimum, productivity to become the main objective. Adverse side-effects of this on livestock, for example mineral disorders, can be overcome and the farmer is learning to recognize soil problems associated with intensive production.

Productivity of grasslands, as determined by dry matter production, is not, therefore, a criterion of much importance to the ecologist, although many still attempt to measure, and assess one ecosystem against another, in terms of production only. Even the agriculturist, to whom production is rightly linked with economics, may be advised to consider whether quality must always be subservient to quantity.

If measures are taken to increase production by chemical additions to a grassland ecosystem, a decline in its botanical interest will result. It follows, therefore, that under present-day agricultural policies there is likely to be a wide difference of objective between the agriculturist and wildlife conservationist. In practice this means that on grassland reserves, managed primarily for their biological interest, the crops (if taken), whether in the form of hay or live-weight gain of stock, must be regarded as of secondary importance to the biological objectives.

The effect of soils, water, geography and management, and their interactions, has resulted in a wide variety of lowland grassland types. Of the 1500 species of flowering plants in Britain, over 500 are associated with these grasslands and 400 or so are most frequently found in this habitat. Diversity and richness are discussed elsewhere but it is pertinent to recall here that some grasslands are as rich, in species per unit area, as any other vegetation formation. A metre square of Wiltshire downland or a dale's meadow may contain up to 40 species of flowering plants.

Festuca rubra is the one constant species of most lowland grasslands which are not significantly 'improved' for agricultural use. Its degree of dominance depends on soil fertility, being more pronounced on the poorer soils. Grasses which have a high degree of constancy are *Agrostis stolonifera*, *Briza media* and *Dactylis glomerata*. Certain grasses are characteristic of particular soil types such as *Zerna erecta* on chalk, *Brachypodium pinnatum* on Oolitic limestone, *Alopecurus pratensis* on clay loams and *Sesleria albicans* on Carboniferous limestone. Other grasses show a relationship to water table and soil moisture – *Poa pratensis* on drier soils, *P. trivialis* on damp soils and *Glyceria fluitans* on wet soils.

Interesting geographical relationships are shown by *Cirsium heterophyllum* and *C. dissectum*, the former only occurring in the north and the latter in the south. This north–south relationship can be carried further in a comparison of the north and south hay meadows. In the former are found *Geranium sylvaticum, Trollius europaeus* and *C. heterophyllum* whilst in the south they are 'replaced' by *Serratula tinctoria, Betonica officinalis, Fritillaria meleagris* and *Orchis morio*. An east–west factor is less evident, but *Carex humilis* is only found on the western chalk and *Trifolium ochroleucon* only occurs in meadows and roadside verges on the clay soils of eastern England.

5 Zoological characteristics and interest of the grassland ecosystem

The Soil Fauna

For convenience two components of the fauna of grassland may be recognized, the animals of the *soil* and those of the grassland proper. This is to a large extent an arbitrary division, for a number of reasons. First, the animals of the *litter*, or surface layer of dead and decomposing plant material, may be considered as being, in a functional sense, part of the soil ecosystem. Indeed, it is misleading to separate the soil from the grassland, as both are mutually dependent. It is the decomposition of plant and animal material by the saprophagous fauna and flora of the litter which produces the organic matter present in the soil as humus. Secondly, the separation of soil and grassland animals is arbitrary because it cuts across the usual vertical divisions used by soil zoologists, who use the terms epigeal zone to denote the grassland vegetation, hemiedaphic zone to denote the litter and surface layer of the soil, and euedaphic zone to distinguish the deeper mineral part of the soil (e.g. Wallwork, 1967). Thirdly, a large number of grassland animals inhabit the soil at particular stages of their life-histories. Various species of Lepidoptera, Coleoptera, Hymenoptera–Symphyta and Diptera feed in the soil or pupate there, while animals such as aculeate Hymenoptera and geotrupid Coleoptera inhabit burrows in the soil.

Soil zoology tends to be a specialized subject for study, and is clearly an important one from the practical point of view of agriculture and maintaining soil fertility. Considerable importance is attached to extraction and sampling techniques because the animals, which are mostly very small, must be removed from the soil matrix before they can be identified and counted (Kevan, 1955; Murphy, 1962). The groups of animals discussed by Burges and Raw (1967) may be taken as an indication of their importance

92

Table 5.1 Estimates of densities of Collembola populations on grasslands and related biotopes (*from* Hale, 1967)

Author	Broad biotope type	Vegetation/soil type	Density: thousands/m²
Glasgow	grasslands	*Dactylis*	27·0 (mean for year)
Salt et al.		pasture	43·0
Weis-Fogh		sandy soil	8·5
Schaller		limestone grassland	15·9 and 25·0
Dhillon & Gibson		loam on boulder clay	33·0 (mean for year, 1960)
Hale		limestone grassland	53·0 (mean for year, 1961)
			42·0 (mean for year)
		alluvial grassland	44·0 (mean for year)
		Juncus squarrosus	21·0 (mean for year)
Strenzke	fen	*Phragmites*	20·0
Macfadyen		*Molinia*	25·0 (mean for year)
		Deschampsia	24·0 (mean for year)
		Juncus subnodulosus	7·2 (mean for year)
Hale	moorland	*Calluna* litter	35·0 (mean for year)
Milne		*Pteridium*	15·5 (mean for year, 1956)
			17·7 (mean for year, 1957)

in soil biology; they are Protozoa; Nematoda; Enchytraeidae and Lumbricidae (Oligochaeta); Arthropoda (Acarina and Collembola being treated separately) and Mollusca. The numbers of soil animals are immense. In a classic paper Salt *et al.* (1948) recorded $263·6 \times 10^3$ arthropods per m^2 of pasture soil, and estimated that the actual total present was at least 346×10^3. As many as 20 million nematodes per m^2 of grassland have been recorded by Nielsen (1967).

In most studies of the whole fauna of grasslands Acarina and Collembola are found to be the most numerous of the macroscopic animals. For instance, in a study using a vacuum sampling apparatus Southwood and Van Emden (1967) recorded mean numbers of 2358 and 1652 animals per m^2 on cut and uncut grassland of neutral to acid type at Silwood, Berkshire. Acarina and Collembola together accounted for 73% and 66% of each total respectively, with the mites 2·1 and 1·3 times as abundant as the springtails. Sheals (1957) found that Collembola and mites accounted for 39% and 58% of the total arthropod fauna respectively in the soil of 'old grassland' near Glasgow. Collembola appear to feed almost entirely on soil fungi and detritus, with the smaller species feeding directly on particles of humus (Hale, 1967). Estimates of population densities on grassland (Table 5.1) show that the figures obtained by Southwood and Van Emden for *total* faunas are very low, the differences presumably being a measure of the Collembola actually present *in* the soil. Because of their abundance both Acarina and Collembola are potentially important sources of prey to other animals, not only of the soil but also those of the vegetation (epigeal) zone as well. Collembola may be particularly important because they often have several generations in a year; the voltinism of Acarina is more variable. Because of their small size, animals of the two groups contribute relatively little to the total biomass of grassland animals, although they are so numerous.

The Influence of Microclimate

Grassland is maintained in a state of structural immaturity by the continual processes of management but, despite this, the microclimate is far from uniform. Partly this is because even the relatively low degree of structural differences within grassland is microclimatologically important, partly because differences of aspect and topography are also important. The differences of insolation (solar radiation) falling on even quite shallow slopes of different aspect are quite dramatic (Geiger, 1965). The combination of such factors as these with soil type may be important. Thus Southwood (1957) has suggested that chalk and sandy soils, which warm up rapidly, support species of Heteroptera which are thermophilous and on the edge of their range in Britain and which cannot persist on colder soils such as clays. It is well-known that mosses grow more luxuriantly on north-facing grassland slopes than on those with a southern aspect, this probably affects not only species such as the tingid bug *Acalypta parvula* which feeds on mosses, but also insects which overwinter among mosses.

Unfortunately there has been little comprehensive investigation of grassland microclimate as it affects animals, although Whitman (1969) has reviewed studies on its effects on plants. An important paper is that of Waterhouse (1955). This shows that in grass crops litter is important in maintaining cool, humid conditions near the ground. In grassland the speed and drying power of the air are greatly reduced in passing from just above the crop to the area just below the top of the vegetation. Humidity appears to be critical for many species of grassland animals. Collembola in particular are limited in distribution by humidity (Sheals, 1957; Hale, 1967) and woodlice also seem to be dependent on the relatively high humidity in longer grasslands. Sutton (1968) showed that populations of *Trichoniscus pusillus* were very much affected by drought, although there was no reaction by another species of woodlouse, *Philoscia muscorum*, which seems to be better adapted than most species to slightly drier grasslands. Long grasslands are almost certainly more equable than short ones. This may be particularly important for overwintering insects, such as beetles. Thus Freeman (1965) has shown that the weevil *Apion dichroum* cannot survive during winter in the clover fields where it breeds (in summer), but must migrate to woodland litter or tall grassland to overwinter successfully. The numbers of beetles in samples collected from ungrazed chalk grassland by Morris (1968) were nearly three times as great as in similar samples taken on grazed grassland, averaged over a year, but in summer the numbers were about equal.

As grasslands are stable only in a dynamic sense the characteristic animals are ones which adapt readily to change, in particular by being mobile and able to colonize new areas rapidly and effectively. An extreme example is the cicadellid leafhopper *Macrosteles laevis*. Andrzejewska (1962) has shown that this species responds to almost any kind of 'unsettlement' in natural meadows and is therefore predominant in the associations of Auchenorhyncha in frequently managed areas, particularly cultivated meadows. A different effect was postulated by Morris (1974), who recorded the leaf hopper *Notus flavipennis* from 34% of samples taken on grasslands in the Burren, Co. Clare, Ireland. It was not recorded at all in a more extensive survey of English calcareous grasslands (Morris, 1971b). *N. flavipennis* feeds on *Carex* spp., which grow on both dry calcareous grasslands and in marshes. Morris suggested that in areas of high rainfall such as the Burren the leaf hopper could survive on calcareous grasslands as well as in marshes.

A feature which is common to grassland and other types of vegetation is the base status of the soil. The amount of available calcium carbonate is a direct factor in determining the presence and abundance of many terrestrial gastropods (snails), for the obvious reason that the shells of snails are composed of calcium carbonate. In an early paper Boycott (1934) was able to classify the British gastropods into groups, not only on the basis of their requirements for lime but also on their degree of tolerance to dry and humid conditions. A number of species are considered to be obligate calcicoles while others are more tolerant of relative lime-deficiency.

The Plant as a Source of Food

Most, if not all, phytophagous animals display some preferences about what they eat. There are, however, very great differences from species to species, from the very polyphagous to the strictly monophagous. A primary consideration determining whether a species of stenophagous animal occurs on a particular site is the presence or absence of its food-plant. However, other environmental factors may be of importance and many examples are known of localities where a food-plant is present but the associated insect species is absent.

Probably because grasses are, on the whole, more difficult to identify than dicotyledonous plants the host species of many grass-feeding animals are not known with certainty. Auchenorhyncha certainly display some preferences in the grasses they attack. For instance *Mocydia crocea* is probably associated with *Zerna erecta*, *Brachypodium*, *Dactylis* and other coarse grasses (Le Quesne, 1969), while a number of species has recently been shown to feed on *Holcus* (Waloff and Solomon, 1973). Generally, however, the feeding preferences of Auchenorhyncha associated with grasses remain to be investigated. An interesting group of Heteroptera feeds on grasses, the 'grass mirids', Stenodemini. Most are found in luxuriant grasslands but species of *Teratocoris* are characteristic of marshes and *Trigonotylus psammaecolor* occurs on maritime sand dunes next to the sea where its foodplant, *Agropyron junceiforme*, grows. Its congener, *T. ruficornis*, is characteristic of fairly dry grasslands while species of *Steno-*

dema, Leptopterna dolobrata and *Megaloceraea recticornis* are more often found in damper, lusher grasslands and also occur in woods. The energetics of a population of *L. dolobrata* were examined by McNeil (1971). Because *Notostira elongata* has two generations a year it is said to be more successful than other grass mirids on frequently cut grasslands (Southwood and Leston, 1959).

Many Lepidoptera feed on grasses; 99 species of 'Macrolepidoptera' which feed on Gramineae are listed by Scorer (1913). Most species of the Crambinae, formerly included in the large pyralid genus *Crambus*, are feeders on the roots or basal stems and leaves of grasses. The different species show sharp habitat preferences, each one tending to occur primarily on a particular type of grassland, but the foodplant preferences have not been fully studied (Beirne, 1952). Crambids are important pests of grasses in some parts of North America (Crawford and Harwood, 1964). The larvae of *Elachista* species (Elachistidae) are miners in the leaves of grasses and species of *Luzula, Carex* or *Eriophorum* and related genera (Ford, 1949). Thirty-six species are listed as British, some showing strict monophagous feeding habits. An economically important group of insects feeding on grasses is the gall-midges (Diptera, Cecidomyiidae). Barnes (1946) gives an account particularly of the seed-feeding species, but also lists the stem-feeders. Gall-midges appear to be more host-specific than most other grass-feeding insects, but this is probably because more accurate records of host preferences have been kept. There is no doubt that much more work remains to be done on grassland Cecidomyiidae, particularly on species feeding on grasses of no economic importance. An interesting group of 'parasitic' Hymenoptera feeding internally in grasses, species of *Tetramesa* (Eurytomidae), is described by Claridge (1961). Most species are strictly monophagous and many have specific hymenopterous parasites of the genus *Eurytoma* (Claridge, 1959). A large number of sawflies (Hymenoptera, Symphyta) feed on grasses. One of the important groups is the Cephidae (*Cephus, Trachelus* and *Calameuta*), which are stem-borers and include the well-known wheat stem sawfly, *Cephus pygmaeus* (Benson, 1951). Several Selandriine genera feed on grasses and related plants, species of *Dolerus* being the most numerous (Benson, 1952), and the same is true of some species of *Pachynematus* (Benson, 1958). Southwood and Jepson (1962) studied stem-boring Diptera of grassland in Berkshire. Most species had a wide spectrum of gramineous hosts, but some, such as *Meromyza variegata* (Chloropidae), were monophagous, in this particular case on *Dactylis glomerata*. A rather similar situation is presented by another family of Diptera, the Agromyzidae, whose hosts are given in

Table 5.2 Species of British *Agromyzidae* (Diptera) recorded from different grass hosts (*after* Spencer, 1972)

	Agromyza albipennis	*alunulata*	*ambigua*	*Cerodontha bromi*	*cinerascens*	*hendeli*	*intermittens*	*lucida*	*mobilis*	*nigrella*	*nigripes*	*nigrociliata*	*phragmitidis*	*rondensis*	*Cerodontha crassiseta*	*denticornis*	*deschampsiae*	*flavocingulata*	*fulcripes*	*hennigi*	*incisa*	*lateralis*	*muscina*	*Liriomyza phalaridis*	*phragmitidis*	*pygmaea*	*superciliosa*	*Liriomyza flaveola*	*pedestris*	*Phytomyza phryne*	*pusio*	*Phytomyza milii*	*nigra*	*Pseudonapomyza atra*
Gramineae *Agropyron caninum*												+				+		+,+				+					+							
repens												+				+		+									+							
Agrostis stolonifera																																		
Alopecurus pratensis																																		
Ammophila arenaria																																		
Arrhenatherum elatius																														+		+[1]	+	
Avena sativa	+		+										+	+		+									+	+	+						+	+
Brachypodium sylvaticum														+							+													
Bromus arvensis														+																				
villosus																																		
spp.																																		
Calamagrostis canescens																			+	+			+[2]					+					+	
epigejos																																		
Ceratochloa unioloides																																		
Dactylis glomerata									+					+				+								+		+					+	
Deschampsia cespitosa																					+													
flexuosa																+	+									++			+				+	+

98

Festuca gigantea
pratensis
rubra
spp.
Glyceria fluitans
maxima
Hierochloe odorata
Holcus lanatus
mollis
Hordeum murinum
vulgare
Lolium perenne
Milium effusum
Molinia caerulea
Phalaris arundinacea
Phleum spp.
Phragmites communis
Poa annua
compressa
trivialis
spp.
Secale cereale
Setaria spp.
Trisetum flavescens
Triticum aestivum

1 *Arrhenatherum* sp.
2 and *Calamagrostis* sp.
3 *Hierochloe* sp.

detail by Spencer (1972). All the grass-feeding species are leafminers. Some are very polyphagous, such as *Agromyza nigrella* and *Phytomyza nigra*, but others are known from only one host (Table 5.2).

The main groups of plant-feeding invertebrates with some stenophagous habits are listed in Table 5.3. Not all the species inhabit grassland, but a

Table 5.3 The main groups of phytophagous invertebrates which have some degree of stenophagous habits, with approximate numbers of British species

	Group			No. species
Acari	Eriophyidae			210*
Insecta	Heteroptera			500
	Homoptera	Auchenorhyncha		350
		Sternorhyncha	Psylloidea	106
			Aleuroidea	24
			Aphidoidea	725
			Coccoidea	300
	Thysanoptera	(some)		160
	Lepidoptera			2400
	Coleoptera	Nitidulidae (some, esp. *Meligethes*)		113
		Phalacridae		22
		Chrysomeloidea		320
		Curculionoidea		530
	Hymenoptera	Symphyta		540
		Cynipoidea (some)		300
		Chalcidoidea (few)		1700
	Diptera	Cecidomyiidae (some)		650
		Trypetidae		92
		Agromyzidae		250
		Chloropidae		125
		Muscidae–Anthomyinae (some)		150

* Described species. Probable number very much higher.

high proportion do. It is not possible in the present state of knowledge to give a comprehensive account of grassland phytophagous invertebrates; a selection of species associated with particular chalk grassland plants was given by Duffey and Morris (1966). Some groups of closely related species show a wide difference in food-plant associations. The British species of *Apion* (Coleoptera, Apionidae) particularly associated with grassland are listed with their food-plants (Table 5.4); these are principally Leguminosae, but include many other plants. Similar lists could be drawn up for *Meligethes* (Coleoptera, Nitidulidae), Halticinae (Coleoptera) or Trypetidae (Diptera), for example. Insects and other invertebrates tend to feed on common and abundant species of plants, although those with local

Table 5.4 Species of *Apion* (Coleoptera) associated with grassland

Species	Plant-host	Structure attacked	Type of grassland inhabited
curtirostre Germ.	*Rumex acetosa*	stems	neutral
haematodes Kirby	*Rumex acetosella*	leaf-petioles	acid
millum Bach	*Prunella vulgaris*	?	mainly calciphilous
flavimanum Gyll.	*Origanum vulgare*	?	calciphilous
vicinum Kirby	*Mentha* spp.	gall-maker on stems	mainly wet meadows
atomarium Kirby	*Thymus* spp.	gall-maker on stems	mainly calciphilous
seniculus Kirby	*Trifolium hybridum* (mainly)	stems	calciphilous and neutral
stolidum Germ.	*Chrysanthemum leucanthemum*	root-stock	calciphilous
onopordi Kirby	*Centaurea* and *Cirsium* spp. etc.	stems	calciphilous and neutral
carduorum Kirby	*Cirsium arvense*	stems	calciphilous and neutral
meliloti Kirby	*Melilotus* spp.	stems	disturbed
waltoni Steph.	*Hippocrepis comosa*	pods	calciphilous, esp. chalk
loti Kirby	*Lotus corniculatus*	stems	calciphilous and neutral
tenue Kirby	*Medicago* spp.	vegetative buds	most
pisi F.	*Medicago* spp., ? other Papilionaceae	stems	most
aethiops Herbst	*Vicia* spp.	?	mostly wet meadows and fens
spencii Kirby	*Vicia* spp.	pods	mainly calciphilous
ononis Kirby	*Ononis* spp.	pods	most
craccae (L.)	*Vicia* spp.	stems	most
virens Herbst	*Trifolium repens*	flower heads	most
dichroum Bedel	*Trifolium repens*	? flower heads	most
nigritarse Kirby	*Trifolium* spp.	flower heads	most
aestivum Germ.	*Trifolium pratense*	flower heads	most
apricans Herbst	*Trifolium pratense*	flower heads	most
assimile Kirby	*Trifolium pratense*		most
ononicola Bach	*Ononis* spp.	pods	calciphilous
dissimile Germ.	*Trifolium arvense*	seed-heads	acid-neutral

distributions are often favoured. These are often calcicolous species; thus *Hippocrepis comosa* is well known as the food-plant of the chalkhill blue and adonis blue butterflies, and is also the food-plant of *Apion waltoni* (Table 5.4). Another example is the calcicole plant *Thesium humifusum*, which is the food-plant of a conspicuous heteropterous bug, *Sehirus dubius*.

The specific parasites of phytophagous animals restricted to one or a few food-plants are conveniently considered in relation to their hosts. In a study of the inhabitants of spear thistle heads (*Cirsium vulgare*), Redfern (1968) found three parasites specifically attacking the common trypetid fly *Urophora stylata*. Many other phytophagous insects were feeding in or on the seed-heads. The knapweed gall-fly, *Urophora jaceana*, is the centre of an even more complex food-web. Varley (1947) lists 24 species of insect from one site as being associated with the seed-heads of knapweed (*Centaurea nigra*), either as phytophagous species or their parasites and predators (one case only). As will be seen later, knapweed supports many insect feeders on its leaves and in its stems, so that the total number of species associated either directly or indirectly with the plant may well reach 50 or more when all the relationships have been determined. Not all plant species are attacked by as many species as is knapweed, but the overall complexity of grassland communities of animals is clearly very great.

The structure of individual species of plants is often of the greatest importance to the associated fauna, particularly the feeders in specialized aerial and reproductive structures. The importance of this factor in the management of grassland for nature conservation has been widely recognized (Chapter 7). Phytophagous animals feed on every part of a plant. Thus the roots are attacked by the larvae of Lepidoptera and Coleoptera, and the root nodules of Leguminosae by larvae of *Sitona* (Coleoptera) and *Micropeza* (= *Tylos*) (Diptera). Stems are the habitat of many species, particularly Diptera, Coleoptera and Lepidoptera; the feeding site may be invisible from outside, or a gall may be formed. Galls may also be formed on similar parts of plants, such as runners, petioles and midribs of leaves. Stems may be important for oviposition rather than as feeding sites. Following Kullenberg's observation (1946) that the bug *Leptopterna ferrugata* oviposits in the stems of grasses below the inflorescence, Morris (1967a) found that more larvae of this species were to be found on ungrazed grassland plots than on similar, but grazed, plots.

Leaves provide the food for a great many phytophagous animals. Feeding may be external, or internal, when the animals usually form characteristic 'mines' in the leaves. Many leaf-mining species have been studied in detail; much of the work is summarized by Hering (1951). Leaf-mining

Coleophoridae (Lepidoptera) in most cases form a 'case' usually made of leaf fragments in which the larva lives. Although some animals, usually the smaller larvae of Lepidoptera, do not mine the leaves internally, they give an impression of so doing by forming 'windows' in the leaves, which they do by removing all the leaf tissue except the upper epidermis. Caterpillars of Lepidoptera and sawflies, and beetles in the adult and larval stage, usually bite fragments off the leaf lamina. Insects with suctional mouthparts, such as Heteroptera and Auchenorhyncha, cause characteristic feeding marks in the lamina, while aphids may produce curling or other kinds of distortion of the leaf-blade.

Morris (1969a) suggests that grassland animals which are feeders on leaves may often be less 'at risk' on intensively managed sites than species which feed on other, more specialized, plant structures. Although no doubt several factors are involved in the occurrence of such species on sites which are intensively managed, many plants respond to repeated defoliation by producing rosette leaves (Norman, 1960; Wells, 1965), thus providing a continuous supply of food for leaf-feeding species. Dempster (1971a, b) has even shown that defoliation of ragwort, *Senecio jacobaea*, by the cinnabar moth, *Tyria jacobaeae*, may result in an increase of the plant population by regeneration from root-buds, in a wet summer. Fleabeetles (Halticinae), most of which are foliage feeders, are often observed to be particularly abundant on short, well-managed chalk grassland compared with taller vegetation which may be much more suitable for other animals.

Little is known about the species of invertebrate animals which inhabit the vegetative and reproductive buds of grassland herbs. Probably these do not support as many insect species as do the buds of shrubs and forest trees. Weevils of the genus *Anthonomus* feed on larvae in buds and flowers, chiefly of Rosaceae. *A. rubi* attacks herbaceous species such as *Fragaria*, as well as *Rosa*, and is a minor pest of the strawberry in addition to occurring on grassland. *A. brunneipennis* is recorded as attacking many different Rosaceae, but in Britain at least appears to be particularly associated with *Potentilla erecta*. The various species of eriophyid mites, cecidomyiid gall midges, and even the Lepidoptera which feed in buds of grassland plants, are not well-known.

Because flowers and inflorescences grow from buds and develop into fruits and seed-heads it is often difficult to distinguish the different stages, particularly where a feeding larva retards development of the floral organs or fruits. Thus the weevil *Miarus campanulae* normally infests the seed capsules of *Campanula* species, particularly *C. rotundifolia*. Often, however, especially at high population density, the flower-buds are attacked before capsules are formed and a bud-gall or flower-gall develops. Flowers usually have a short life in comparison with buds, fruits and leaves and the thin sepals and petals are in most cases unsuitable for supporting endophagous animals. Flowers are very attractive to insects, however, both nectar and pollen being taken by visiting species. Some species of bees gather pollen predominantly from one species of plant and may thus be considered as stenophagous. Many species of Thysanoptera feed in flowers. Ward (1966) described the biology of five species of thrips which breed in the flowers of *Leontodon hispidus* on chalk grassland in Britain. Flowers are a very ephemeral habitat for larval insects; for instance Morris (1967a) found that the mean length of life of flowers of *Campanula rotundifolia* was $6 \cdot 07 \pm 1 \cdot 15$ days.

Unlike flowers, fruits and seeds often persist for many months in grassland. The seed-heads of *Centaurea nigra* and *C. scabiosa* last throughout the winter. The insects inhabiting seedheads of *C. nigra* at one particular site (Varley, 1947) have already been mentioned. Other species not found at Varley's experimental site could be considered. Each plant supports a species of cynipid gall-wasp which feeds in the achenes of the plant, *Isocolus jaceae* on *C. nigra* and *I. rogenhoferi* on *C. scabiosa*. Two other species of *Isocolus* gall stems and stem-bases of the two plants (Eady and Quinlan, 1963). Seedpods of leguminous plants are particularly liable to attack by insects because of the high protein content of the seeds. Some of the *Apion* species feeding in seedpods are listed in Table 5.4. It is very likely that complex communities living in the seedpods of grassland Leguminosae will be found to occur, similar to that described by Parnell (1962) for the shrub *Sarothamnus scoparius*.

Because most forms of management of grassland remove the aerial and particularly the reproductive parts of plants, the animals which are dependent on these structures usually occur in low numbers on managed grassland. Morris (1967a) describes as examples the weevils *Miarus campanulae* and *Apion loti* which feed on the seeds of *Campanula rotundifolia* and *Lotus corniculatus* respectively. A more recent example is the fauna of *Centaurea nigra* recorded on grazed and ungrazed grassland in Bedfordshire (Morris, 1971b). In 1969 seven species of insect were recorded on *C. nigra* on ungrazed plots but only one of these was recorded, and that doubtfully, from the same plant under grazed conditions. Three species were trypetid flies breeding in the seed-heads and three leaf-feeding Coleoptera and Lepidoptera; a species of aphid was predominantly an external feeder on the stem. A different effect of management was recorded by Morris (1969b) on the bug *Agramma laeta* feeding on *Carex* spp., particularly *C. flacca*. An initial increase in numbers following cessation of grazing was later followed by a rapid decline, possibly as a result of the leaves of the food-plants becoming brown, attenuated and unsuitable as food following the more vigorous growth of competing grasses. It is clear that the structure of individual species of plants in grassland is of very great importance even to those stenophagous animals whose presence or absence in a particular area may be considered to be primarily a matter of whether the food-plant grows there or not.

Grassland Structure and its Influence on the Fauna

Because management changes the structure of grassland the effects of management on the fauna have received considerable attention from ecologists. Structure influences the fauna very profoundly, even when reduction or elimination of aerial parts of individual plant species is not important as with generalized (or at least unspecialized) phytophagous animals and predators. Richards and Waloff (1954) showed that the British Acrididae (grasshoppers) have different habitats which are largely related to structure (see also Elton, 1966; Morris, 1967a). In particular *Chorthippus brunneus* and *C. parallelus* occur in 'short' and 'long' grass respectively. Many grasshoppers oviposit in compacted soil (Richards and Waloff, 1954; Dempster, 1963); this is another structural component of grassland which is important for some species. Rather little is known about how habitat requirements change in grassland during the course of development although it is suspected that the different stages in the life history of at least some species do have different requirements.

Most groups of animals occur more abundantly on ungrazed grassland than on grazed (Morris, 1968). Some of these animals respond, at least in part, to the microclimatic changes which occur when grazed turf grows up or ungrazed grassland is once more grazed – for instance the woodlice. Others may be affected by the larger quantities of food which are available in ungrazed grassland, whether this be fresh plant material, dead vegetation or prey animals. Cover is also an important requirement for many animals, as pointed out by Elton (1966). The leafhoppers (Auchenorhyncha) form a parallel case to the Acrididae in that some species may be more abundant on ungrazed grassland while others occur predominantly on grazed turf, although the former are much more numerous (Morris, 1971a). This is true not only of the number of species occurring, but also of the number of individuals, while diversity of leafhopper samples is always higher on ungrazed land. In extensive studies of the leafhopper faunas of limestone grassland sites, Morris (1971b, 1974) found that these facts were statistically true for the range of structural types examined. On average, the taller a grassland is, the more species and individuals will it support per unit area and the greater will be the value of an index of diversity calculated on the basis of samples taken from areas of grassland of the same size. Andrzejewska (1965) has shown that in meadows Auchenorhyncha species are stratified, or organized in vertical components of the grassland. Whittaker (1969) was able to assign the species of Auchenorhyncha taken on coral rag limestone grassland near Oxford to different 'habitat types' present in the locality. Morris (1971a) found that there was generally good agreement between Whittaker's assignments and his own observations. The Heteroptera of limestone grasslands show similar responses to height of the vegetation as do the Auchenorhyncha. In general, so do the weevils (Coleoptera, Curculionoidea) but here there is a suggestion that numbers of species and individuals fall off in the very tallest grasslands (30 cm +) (Morris, 1971b).

As well as being greatly affected by management and its intensity, leafhopper faunas, and no doubt other invertebrate populations, are also influenced by the timing of management. Morris (1973) studied the Heteroptera and Auchenorhyncha of chalk grassland plots at Aston Rowant National Nature Reserve which were grazed for three months in autumn, winter, spring and summer. Numbers of adult individuals and species of both groups tended to be higher on plots grazed in autumn and winter than on those treated in spring and summer, largely because the insects became adult during the latter periods. One species, *Zygina scutellaris*, showed a positive response to spring grazing, but was exceptional. Generally, phenology of habitat components such as food and cover is important to grassland animals.

Southwood and Van Emden (1967) studied the differences between cut and uncut acid grassland and found that the former supported more invertebrate animals than the latter. Recent unpublished work on the mowing of limestone grassland plots at Castor Hanglands National Nature Reserve, Huntingdon and Peterborough, showed that cut grasslands contained fewer species and individuals of leafhoppers. Recovery of populations following a cut in May was rapid but cutting in July was more permanently damaging. There are thus some controversial features of the faunas of managed grasslands which require resolution. Southwood and Jepson (1962) demonstrated that cut grasslands were more productive of stem-inhabiting Diptera than uncut ones. In general, structure of the grassland appeared to be more important to this group than the species of grass attacked (see p. 102).

Many predatory animals have been shown to require structural components in grassland. One of the most interesting groups in this respect is the Araneae (spiders) studied on limestone grassland by Duffey (1962a, b). Spiders are particularly characteristic of the ground zone (0–6 in above ground level) of grassland. On the limestone grassland site studied by Duffey, 63% of the 141 species recorded were assigned to this 'habitat type'. Open ground, that is the ground zone where there is no taller vegetation, is an important biotope for spiders, both on the chalk (Duffey and Morris, 1966), and Breckland 'grass heaths' (Duffey, 1965), for example. In the grassland field layer (6 in to 6 ft) Duffey (1962b) distinguished four types of spider:

1. Species living permanently on field-layer plants.
2. Species using field-layer plants for construction of egg-cocoons.
3. Species hunting on field-layer plants.
4. Aeronautic species using field-layer plants (as platforms from which to become airborne).

This analysis is particularly interesting because it shows that structural components of grassland can be important to different species for quite different and distinct reasons.

Another characteristic group of predacious invertebrates of grassland is the Nabidae (Heteroptera). They appear to be associated mostly with tall vegetation and Fewkes (1961) has shown that they exhibit diel vertical movements in grassland; these are suggested to be microclimatic responses rather than signs of nocturnal activity coupled with diurnal inactivity. The species of Nabidae show interesting patterns of ecological distribution with respect to grassland vegetational types; this may perhaps be interpreted as 'niche differentiation'. Thus *Dolichonabis lineatus* occurs in wet areas of marshes, but *D. limbatus* has a wider distribution in marshes and damp grasslands. *Nabis ericetorum* is a healthland species. *Nabis ferus* is fully winged and is the commonest nabid to be taken in light-traps (Southwood, 1960); it appears to be the first nabid colonist of grasslands which are allowed to grow up from a grazed condition (Morris, 1968).

So far only the vertical structure of grassland as it affects animals has been considered. There are other important features which may be important to animals. When grassland is well grazed one effect of the grazing animal is to compact the soil (Arnold, 1964). This has the effect of keeping the soil surface reasonably level. In ungrazed grassland tussocks are formed and the grassland becomes uneven. The formation of tussocks seems to be primarily dependent on the growth form of grass species, particularly the coarser ones such as *Dactylis, Brachypodium* and *Zerna*.

A contributory factor appears to be the activity of small mammals (Elton, 1966). Luff (1965) showed that the microclimate of *Dactylis glomerata* tussocks was different in several respects from the microclimate of the shorter grassland in which the tussocks occurred. He demonstrated (Luff, 1966) that the tussocks provided habitats (microhabitats) for some species of Coleoptera; they also appear to be important overwintering sites for other species of beetles and other animals. Morris (1968) found that ungrazed grassland supported more species and individuals of Coleoptera in winter than in summer, relative to near-by grazed grassland.

Other structural features of grassland in the horizontal plane include ants' nests, molehills and the nests of mammals and birds. Such structures may support animals other than those which built or made them, and their parasites. They are termed animal artefacts by Elton (1966). A popular account of Coleoptera associated with moles' nests is given by Sharp (1915). Donisthorpe (1927) describes the inquilines of British ants; species of *Formica* have the largest number of associated invertebrates and as nests of species of this genus do not generally occur on grassland the fauna of ants' nests in grassland may be regarded as rather poor. The importance of the large blue butterfly (*Maculinea arion*), which, as a larva, is associated with species of the ant *Myrmica*, will be described later in this chapter. The burrows of mammals are another form of animal artefact which are the specialized habitat of certain invertebrates, notably Coleoptera. Some aspects of the fauna of rabbit burrows are mentioned by Welch (1964).

Many different features may be found in grassland which are not strictly 'structures' of grassland type but which diversify the fauna. Thus isolated trees and small patches of scrub will support invertebrate animals which are not grassland species. Another type of diversifying element in the grass-land scheme is various forms of water body. On dry grasslands which were grazed, particularly, water had to be provided for stock. Such water was usually colonized by plants and animals. Dew ponds are, or were in former times, very characteristic features of the open chalk pastures of southern England. The water-bug fauna, for example, (Macan and MacFadyen, 1941) is more closely related to the grassland ecosystem than would be thought from casual consideration because the contamination of dew ponds by the dung and urine of stock is responsible for the presence of such species as *Sigara lateralis*.

Meadows may be much more usually characterized by the presence of bodies of water, either as streams or rivers flowing through or beside them, or as actual water courses used to 'irrigate' the meadows (Sheail, 1971a). At the edges of all such water bodies the grassland may form part of an 'aquatic-terrestrial' transition zone which will have a sharp boundary if the water course has been canalized but in more natural conditions may form an extensive marsh. Transition zones of this latter type often support large numbers of animals which are not part of either the grassland or aquatic ecosystem, for instance phytophagous species associated with such plants as *Scrophularia aquatica* and *Mentha aquatica*. Also present in most cases are those animals with a free-living aerial adult stage but aquatic larvae. Aquatic-terrestrial transition zones are often characterized by large numbers of these animals such as Trichoptera, Odonata, and Megaloptera.

On a rather larger scale lakes and ponds may be considered as integral parts of the grassland biome. An account of this aspect of the ecology of grasslands is given by LaVelle *et al.* (1969).

Species of Conservation Importance

Although all grassland animals are of interest to the zoologist and ecologist most are abundant and widely distributed and call for no special attention as far as their conservation is concerned. On the other hand there are many species which are uncommon or rare or which have very restricted habitat requirements. Of these the most important may be considered to be those which are attractive and distinctive enough to be observed by the general naturalist and to be potentially at risk through unscrupulous collecting as well as from environmental changes and destruction of habitat. Most of these species are Lepidoptera and many are butterflies. In Britain, distribution maps of these insects, the most frequently collected and observed, show that many species are extremely localized (Heath, 1970). On the other hand, only three species of the 54 or so resident in Britain do not occur on National Nature Reserves (Morris, 1967b). One of these three is a grassland species and the most vulnerable of all, the large blue butterfly, *Maculinea arion* (Plate 10b). Usher (1973) stated that 'it is certain that the collection of large numbers of specimens has resulted in the elimination of this species from many of its former inland haunts'. No evidence is put forward for this view, which is contrary to the careful analysis of the available information made by Spooner (1963). He showed that this butterfly, now restricted to a very few sites in south-west England, was once widely distributed over the whole of southern England. Its decline can probably be attributed partly to changing agricultural practice, particularly, of course, to ploughing up of former rough pastures, and partly to adverse weather. There is an increasing body of evidence which suggests that sunshine is a much more important factor than was at one time thought in the ecology of many butterfly species, including *M. arion*. No intensive work has yet been done on the dynamics of populations of the large blue, but Spooner suggests that a balance needs to be struck between the over-management of a site and its neglect with subsequent swamping of *Thymus*, the food-plant of the species before it enters ants' nests and feeds on the larvae. While this is perhaps a truism it is certain that the management of nature reserves for populations of butterflies could be complex.

Observations suggest that patchy vegetation such as is more often produced by cattle grazing than by stocking with sheep is often particularly attractive to butterflies such as the small blue, *Cupido minimus* (Plate 10a) and the chalkhill blue, *Lysandra coridon* (Plate 10c). Both species are characteristic of chalk grassland although the small blue also occurs elsewhere. The foodplants of the two species are *Anthyllis vulneraria* and *Hippocrepis comosa* respectively. Lipscomb and Jackson (1964) were of the opinion that over-grazing was responsible for the decline or destruction of certain populations of butterflies, particularly of *Lysandra bellargus*, the adonis blue (Plate 10d), in Wiltshire. Frazer (1961) recorded a decline to extinction in this species at a site in Kent from 1955 to 1959, but the cause was not ascertained. Although the outlines of life-histories of all British butterflies are well known there have been very few studies of the population dynamics of the different species. Most population studies are made on species of economic importance such as pests and not on rare and local species whose conservation is of importance. Duffey (1968) studied the ecology of the reintroduced large copper *Lycaena dispar batavus*, at Woodwalton Fen, Huntingdonshire. This species is one of fens rather than grassland and although its food-plant, *Rumex hydrolapathum*, often grows in grassland situations it can only regenerate in conditions of open, water-logged ground, that is, a well-formed aquatic-terrestrial transition zone. Duffey concluded that a viable population of *L. dispar* could not be maintained at Woodwalton Fen, because the area available to the butterfly population was too small.

In the absence of information on the natural regulation and control of population numbers of species of high conservation interest, management for the maintenance of these species will have to be based on life-history studies and other data. In some cases the prescriptions are fairly obvious. For instance the attractive little oecophorid moth *Hypercallia citrinalis* breeds with certainty in England only at one spot on the North Downs. This area has become badly overgrown with scrub and measures have been taken by interested entomologists to clear it at the breeding locality. The food-plant is the common grassland herb *Polygala vulgaris.* In the case of other species the type of management required may not be so clear. Two rare species of moth occur (or used to occur) on the chalk grassland reserve at Wye and Crundale. There is no information about the ecological requirements of these species, which are both grass-feeders, although there are indications that their food-plant tolerance is rather wide. *Pachetra sagittigera* (feathered ear) is more definitely associated with grasses than *Siona lineata* (black-veined moth). An example of a species limited in its distribution by its food-plant is the viper's bugloss moth, *Hadena irregularis*, which feeds on *Silene otites* and is confined to grassland sites in the Suffolk Breckland.

Not all species of high conservation interest are Lepidoptera. The Roman snail, *Helix pomatia*, is local in southern England and excites attention because of its size. However, its conservation appears to be limited because the factors affecting its presence and numbers seem to be related to climate, particularly temperature. British grasshoppers of grassland are mostly common and widespread, such as the striped-winged grasshopper *Stenobothrus lineatus* (Plate 11). *Chorthippus vagans* on heaths and *Stethophyma grossum* in bogs, however, are very local. The wartbiter bush-cricket, *Decticus verrucivorus*, is a very rare species which seems to be restricted to calcareous grasslands in the extreme south of England, while the field cricket *Gryllus campestris* is also now one of our rarest Orthoptera. Such very rare species are particularly difficult to study, especially if population dynamics are being considered.

Some species of grassland invertebrates are of conservation importance, most contribute to diversity but are otherwise neutral. At the other extreme a few species are pests. Compared with the pests of arable and horticultural crops the animals attacking grassland have been little studied in Britain, although recently renewed interest is being shown in this field. A full account of grassland pests, whether in Britain or abroad, where they are often of great economic importance, is outside the scope of this book. A useful brief account is given by Moore (1966).

The Vertebrate Fauna

The richest grassland vertebrate faunas are found in the savannas of tropical Africa, and include a complex assemblage of grazers, browsers, and predators. Savanna grassland, with its scattering of bushes, is more productive than pure grassland and in the USA it has been shown that this type of formation may produce five times as much plant material above ground per unit area as does near-by pure grassland. Comparable grassland faunas to those in Africa, but less rich in species, occur in Australia, South America and Southern Asia. Both savanna and tropical grassland have long been subject to fires from natural causes such as lightning and from the activities of early man, with consequent modification of the vegetation structure and the wildlife.

In temperate regions grassland vertebrate faunas are generally much less rich because the vegetation is restricted to a single major stratum. Birdlife tends to be limited in range of species except in the wetter areas, and the mammalian faunas are characterized by small burrowing animals and a few others which are adapted for fast running on the surface. The faunas are limited not only by vegetation structure but also by the lower biomass of plant material available for exploitation: $0.5-3.0$ kg/m^2 compared with $20-60$ kg/m^2 in mature woodlands (Whittaker, 1970). In the latter case the timber is indirectly utilized by vertebrates, particularly birds which prey on insects feeding on both living and dead wood. In Britain grassland climaxes are only well developed above the tree line, and in the lowlands they are very localized on areas such as sea cliffs, dunes and possibly some heathlands, so that few species of vertebrates have been able to establish themselves. For example, in the Scottish Highlands, where there are large areas of upland grassland and moor, golden eagle (*Aquila chrysaëtus*) hunting territories vary between 4000 and 18 000 a. (1620 to 7290 ha) (McVean and Lockie, 1969) and tend to be larger in the west where the prey animals, rabbits, hares and grouse, are less common. In these areas the eagle becomes more dependent on sheep and deer carrion than on live prey so that there is a direct link with farming and game preservation. McVean and Lockie (1969) point out that the margin of safety between survival and death of sheep and lambs each spring in Highland Scotland is a narrow one and each year there are losses from starvation, accident and disease. In Wester Ross, in April 1965 after an average winter, they recorded 15 dead sheep and five dead deer along a 100-mile transect (161 km). Short-tailed voles (*Microtus agrestis*) are sometimes very numerous in upland areas and as they need about 2.1 ft^2 (1951.0 cm^2) of grassland per vole per day to sustain them (Chitty, Pimentel and Krebs, 1968), they may compete with sheep for available herbage.

Rabbits and other small mammals

In lowland Britain before myxomatosis, the rabbit was one of the most important biotic influences maintaining grassland areas, particularly on marginal land, and also during periods of agricultural depression when some pastures ceased to be grazed by sheep and cattle (see also Chapters 1 and 8). Recent studies on the history of the rabbit in Breckland are of special interest in understanding how the status of plants and animals may be changed. Most of this information is derived from Watt (1971) and Crompton (1972) who studied the ecological history of Lakenheath Warren, a large grass heath which, until 1942, extended over 2300 a. (932 ha). During World War II it was reduced to about 1200 a. (486 ha) but until then it was the only Warren to survive in Breckland out of the many which were established during the second half of the thirteenth century.

During the 600 years the Warren was managed for rabbits, a boundary bank was maintained to keep the animals inside and also to provide a site for trapping them between September and March. The density of rabbits appears to have increased steadily during the Warren's history and in the eighteenth century it was estimated at 15–18/a. (37·0–44·4/ha), the highest figures recorded. The numbers declined during the nineteenth century as sheep-grazing and management for game birds became more important, so that by the beginning of the twentieth century the rabbit population was probably at the same level as in pre-Tudor times. Between 1915 and 1940 the average number of rabbits killed annually was 1·8/a. (4·4/ha) while on unfenced breck grasslands at Foxholes and Weeting the average was 8/a. (19·8/ha). This considerable difference is thought to have come about because the Warren rabbits bred only once, in the spring, while free-ranging rabbits had an extended breeding period. During World War II rabbits were exterminated on the Warren but began to increase again after 1945 and about 2000 were being killed annually just before the arrival of myxomatosis in 1954.

In the absence of rabbits and other large herbivores, bush growth, particularly seedlings of *Pinus sylvestris*, spread over the Warren so that by 1970 between a third and a half of the area had been colonized. Plate 12 shows the difference between rabbit-grazed and ungrazed grass heath at Weeting. The influence of grazing pressure on the regeneration of woody species is emphasized by the fact that on the whole Warren there is only one tree which pre-dates (by three years) myxomatosis. Certain rare plants known to be preferentially grazed by rabbits have appeared for the first time, or are now more widespread and extending their range: for example, *Deschampsia flexuosa, Helictotrichon pratense, Anthyllis vulneraria, Knautia arvensis* and *Scabiosa columbaria*. On the other hand, elsewhere in Breckland a number of rare plants associated with bare ground or short open vegetation, formerly maintained by the burrowing and scratching of rabbits, are now much less common, particularly *Artemisia campestris, Silene conica, Anisantha tectorum, Phleum arenarium, Veronica triphyllos, V. verna* and *V. praecox*. Many gorse bushes, *Ulex europaeus*, spread over the heathlands in the absence of grazing, but as they grew larger and rabbits again became more numerous, they provided shelter for burrows which were dug around the roots. If this activity is extensive the bush dies and the cycle starts over again. The disappearance of open ground has also resulted in some invertebrate animals, which prefer such conditions, or require soft sand for burrowing, becoming much less common.

It is believed that on calcareous soils a reduction in grazing leads to the spread of aggressive dominant grasses such as *Brachypodium pinnatum* and *Zerna erecta*. Hope-Simpson (1941) has described this process on the South Downs for the period 1920–35. More precise information is available for the spread of *Brachypodium pinnatum* on coral-rag limestone near Oxford. A small area of about 3 a. (1·2 ha) of *Festuca rubra* grassland was heavily grazed by rabbits during the years prior to 1938. With a reduction of rabbits during the War years *B. pinnatum* began to spread from a point in the northern part of the area and by 1946 (Plate 13) about a quarter of the area was covered. In 1953 (Plate 14) the spread of this grass had extended to about a third of the area, although the boundaries between it and the fine-leaved turf were still very sharp due to rabbit grazing. After the virtual disappearance of the rabbit in 1954 the rate of change accelerated (Plate 15 shows the situation in 1960) and by 1969 *B. pinnatum* had invaded the whole of the *Festuca* area. In this particular case the rabbit was the only large herbivore having a significant effect on the vegetation but there is some evidence to suggest that a combination of trampling by sheep or cattle, together with rabbit grazing, may be effective in controlling this aggressive grass.

The relatively tall growth of *B. pinnatum* leads to the accumulation over the ground of a fairly dense layer of leaf litter which breaks down rather slowly, probably because the high silica content makes it unpalatable to saprophagous invertebrates. At the same time this layer provides excellent cover for the short-tailed vole which also eats the leaves and stem bases of the living grass. Godfrey (1953, 1955) analysed 119 groups of faecal pellets of voles from limestone grassland near Oxford, where vole densities of 90–120/a. (222–296/ha) were recorded. She found that *B. pinnatum* was the most frequently recorded food material and in one area it occurred in 93% of the samples. Chitty, Pimentel and Krebs (1968) found that the short-tailed vole could live and grow on a diet of this grass alone, although it did better on a balanced 'laboratory' diet. By feeding on grass stem bases and constructing numerous runways, vole activity accentuates the patchiness and tussocky formation in grasslands (see page 105 for the influence this has on the invertebrate fauna).

Cyclic fluctuations in the numbers of voles, and the problem of obtaining a random sample by trapping, make it particularly difficult to measure accurately population density. Chitty *et al.* (1968) recorded spring numbers of 30 *Microtus* and 20 *Clethrionomys glareolus* (bank vole) per acre (0·4 ha) in ungrazed limestone grassland in Berkshire, while Hansson (1968), working on the small mammals on an abandoned field and a pasture in southern Sweden, recorded from 5 to 60/ha for *Microtus agrestis*, 5/ha for the wood mouse (*Apodemus sylvaticus*) and 3 to 10/ha for the common shrew (*Sorex araneus*). Other estimates for *M. agrestis*, quoted by Hansson from other sources, vary from 70/ha in permanent grassland to 750/ha in a young conifer plantation. He also found that *A. sylvaticus*, in contrast to *M. agrestis*, tended to leave the open fields and move into adjacent forests

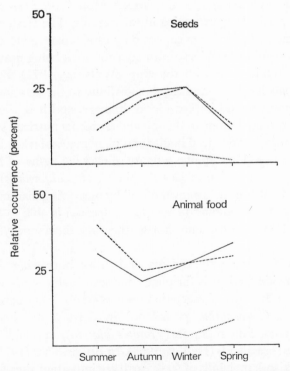

Fig. 5.1 Relative occurrence of seed and animal food taken by the short-tailed vole, *Microtus agrestis*, the bank vole, *Clethrionomys glareolus* and the wood mouse, *Apodemus sylvaticus*. (*After* Hansson, 1971.)

during the autumn. This seems to be because the wood mouse and the bank vole are mainly granivorous at certain times of the year (Fig. 5.1), and not so successfully adapted to a diet of plant material as is the short-tailed vole (Hansson, 1971). However, Watts (1968) found that although the bank vole ate many seeds in its normal diet it was able to survive without them and that during the winter a surprisingly high proportion (up to 50%) of the food intake consisted of dead leaves, those of woody species being preferred to herbs. Hansson showed that rodent herbivores needed more water and had a more even diel rhythm of food intake than the night-active granivores. In the presence of a rich source of tree seed, granivorous species (*A. sylvaticus* and *C. glareolus*) strengthened the period of reproduction, had higher body weights and increased the population density during the same autumn and following spring.

The burrowing and scratching activity of small mammals, particularly rabbits, maintains a significant proportion of bare ground in areas with dry, light soils. On heavier soil types burrowing probably has little effect apart from being a nuisance to the farmer. Mole (*Talpa europaea*) activity may also be particularly damaging in this respect. The fresh earth thrown up in high mounds, for instance, not only causes damage to the knives of mowing and reaping machines, but may ruin grass being grown for silage purposes by mixing soil with the crop (Mellanby, 1971). Burrowing by small mammals has been claimed to contribute to the aeration of the soil, to increase porosity and improve local drainage, and it is also said that it helps to incorporate humus in the soil and enrich its nutrient status by the addition of faeces, urine, dead bodies and plant material taken in for nesting purposes. De Vos (1969) quotes a figure of over 6·6 kg/ha of faecal pellets deposited by a population of jack rabbits (*Lepus californicus*) in Arizona. There are few data on the turnover of soil by burrowing mammals although the American pocket gopher (*Thomomys talpoides*) is able to bring 5 tons of earth/a. (12·6 metric tonnes/ha) to the surface in one year (Ellison, 1946).

Earth mounds made by rodents and other burrowing mammals in grassland provide seed beds for plants and also enable seed-eating species to disseminate the seed of vegetation such as scrub. For example the sand pocket mouse (*Perognathus penicilatus*) in Arizona has been recorded carrying the seed of the mesquite (*Prosopis juliflora*), an established invader of over-grazed grasslands (De Vos, 1969). In Britain Watt (1971) has shown that, in Breckland, molehills of fresh earth are important sites for maintaining a cycle of plant succession beginning with annual herbs, followed by perennial herbs and finally perennial grasses. Elton (1966) describes the succession of vegetation on molehills in French pastures. In the first year six pasture plants and six invading species were recorded. By the second year, 21 pasture species were present together with eight pioneer species, while in the third year the molehill vegetation closely resembled that of the surrounding pasture.

The bird life

The common birds of British grasslands are known to most people who walk in the countryside. In lowland regions the skylark (*Alauda arvensis*) and the lapwing (*Vanellus vanellus*) are widespread and common and in the uplands the curlew (*Numenius arquata*) and meadow pipit (*Anthus pratensis*) are probably the most characteristic birds. Other species, generally less numerous and sometimes rather local in distribution, will be known to those familiar with their breeding haunts, for example, whinchat (*Saxicola rubetra*), yellow wagtail (*Montacilla flava*), corn bunting (*Emberiza calandra*) and the redshank (*Tringa totanus*) in the lusher meadows (Plate 16), especially where conditions are moist as in river valleys, the grass-hopper warbler (*Locustella naevia*) in rank grassland where there is plenty of cover (Plate 17) and the woodlark (*Lullula arborea*) and stone curlew (*Burhinus oedicnemus*) on the stony Breckland grass heaths and on the chalk downs. Some of Britain's rarest breeding birds are confined to grass-lands, notably the black-tailed godwit (*Limosa limosa*) and ruff (*Philo-machus pugnax*) in the wet meadows of the East Anglian Washlands. Grass-lands and heaths, especially in upland regions, are important as hunting territories for raptorial birds such as the short-eared owl (*Asio flammeus*), montagu's (*Circus pygarus*) and hen harriers (*C. cyaneus*) and the merlin (*Falco columbarius*). Where grasslands are invaded by bushes and trees many other species occur, particularly the linnet (*Acanthis cannabina*), yellow bunting (*Emberiza citrinella*), willow warbler (*Phylloscopus trochilus*), whitethroat (*Sylvia communis*), stonechat (*Saxicola torquata*) and red-back-ed shrike (*Lanius collurio*), although the last is largely confined to southern England and has been declining in recent years.

In the winter months the familiar fieldfare (*Turdus pilaris*) and redwing (*Turdus iliacus*) feed in the shorter, grazed, grasslands together with the resident song thrush (*Turdus philomelos*), blackbird (*T. merula*), starling (*Sturnus vulgaris*) and rook (*Corvus frugilegus*), while flocks of golden plover (*Pluvialis apricaria*) visit many lowland areas. In weedy pastures tree sparrows (*Passer montanus*) and bramblings (*Fringilla montifringilla*) search for seeds together with linnets and goldfinches (*Carduelis carduelis*), while grasslands subject to winter flooding attract many surface-feeding duck, geese and swans and the best areas of this type have a very important conservation function as wildfowl refuges.

Although the extensive areas of grassland which survived on the chalk downs up to 30 years ago are now gone or fragmented into small units, some idea of their birdlife can be obtained from the work of Colquhoun and Morley (1941). They made a series of counts on the Berkshire Downs in 1940, each of which covered 26 miles (41·8 km), over a period of 13 hours during the winter, and again in the pre-breeding and breeding seasons. A total of 13 species was recorded, of which the commonest in the winter were the partridge (*Perdix perdix*), rook and starling, in the pre-breeding season the skylark, partridge, lapwing and starling, while the breeding population was dominated by the skylark, meadow pipit, partridge and lapwing. Other breeding birds included the wheatear (*Oenanthe oenanthe*), rook, starling, carrion crow (*Corvus corone*), kestrel (*Falco tinnunculus*) and cuckoo (*Cuculus canorus*). The total population per 100 a. (40·5 ha) was estimated at 14, 35 and 32 respectively for the three periods. This is lower than the density of breeding birds on heaths and moorlands recorded by Lack (1935), who obtained a figure of 67 per 100 a., and much lower than recent estimates for scrub and woodland bird faunas. For example, Murton (1971) estimates that the total bird population per 100 a. of oak woodland in the summer may be in the order of 500 to 550. On 200 a. (81·0 ha) of predominantly arable farmland in Cambridgeshire, Murton recorded 303 birds per 100 a. in the summer and 465 in the winter, illustrating the influence of different types of crops as a food source and hedges and hedgerow trees in providing nest sites in addition to those available on the ground.

Scrubland, which often develops on unused grassland, may have a very high breeding bird population (see Chapter 6 for detailed discussion on scrub faunas). Venables (1939) studied the extensive grasslands of the South Downs a few years before most disappeared under the plough. He found that in his study area some 16 species of birds were recorded in the grassland region but where scrub occurred (hawthorn, juniper, yew) the total increased to 36. The commonest grassland birds were the skylark (10 pairs/mile^2 – 2·59 km^2), meadow pipit (7 pairs/mile^2) and lapwing (5 pairs/mile^2). Where scrub was well-developed the chaffinch, willow warbler, blackbird and robin (*Erithacus rubecula*) were the most numerous species. In Wales, Hope-Jones (1972) made a study of the change in bird numbers and species through a succession of habitats from open gronud to low, medium and high scrub, and mature woodland. The unmber of species recorded was lowest in the open grassland and steadily increased through the various succession stages to a maximum in mature oak woodland. There were six species and 52 pairs/100 a. in the open ground and 31 species and 368 pairs/100 a. in the mature woodland. Fig. 5.2 from Hope-Jones (1972) shows a suggested cycle of succession through the habitats described above together with their wildlife.

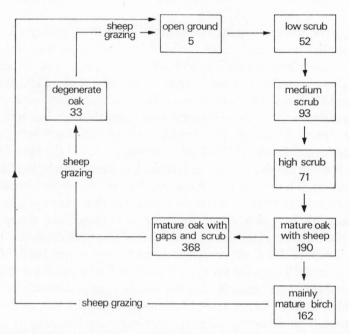

Fig. 5.2 Succession stages in sessile oak (*Quercus petraea*) in western Wales showing the numbers of breeding birds per 100 a. (*After* Hope-Jones, 1972.)

The evidence so far suggests that short grassland characteristic of downs and heaths is generally a rather poor breeding habitat for birds, although it may be important as a source of food for species which nest elsewhere. Birds, such as the starling which nests in holes above the ground and the rook which nests in trees, have bills well-adapted for probing in short, grazed pasture and their diet includes a high proportion of soil invertebrates. These two species, together with the lapwing, were shown by Murton to be the commonest birds in pastures or clover leys on a Cambridgeshire farm in the summer. In the winter the rooks moved away from the pasture to the grain stubble, but in Aberdeen (Dunnet and Patterson, 1968) the preferred animal food of these species (leather-jackets and earthworms) was said to be more abundant in the grasslands during the winter months, and in the same areas in the summer there was a food shortage. The importance of both grassland and stubble to the rook was emphasized by the great increase in the number of rookeries in the South Downs during the 1940s when so much of the grassland was ploughed up for cereal growing. However, an increase in the rook population during this period was reported from several parts of Britain and also in the Netherlands (Murton, 1971).

The type of management applied to grasslands has a significant influence on the species and numbers of birds which breed and feed there, but studies of this phenomenon are very few. The numbers of skylarks and lapwings have been shown to vary in relation to height of sward and both redshanks and black-tailed godwits (*Limosa limosa*) often prefer tussocky cattle-grazed grasslands where there is long grass to hide the nest, but at the same time the close proximity of open, muddy areas and shallow water for feeding. Festetics and Leisler (1968) made a detailed study of the birdlife in the grasslands along the shores of Lake Neusiedl, eastern Austria, in relation to cattle grazing. This type of land use survived in this region because the poor saline soils were of no value for arable farming. However, in recent years vineyards have slowly extended over these grasslands, almost to the water's edge in many places, and the number of cattle fell dramatically. The authors illustrate this trend by reference to an area where herds of about 1000 cattle in 1958 declined to only 320 in 1968. Cattle grazing and trampling maintained a short grass habitat and checked the growth of reed so that areas which were no longer grazed became covered with tall, dense stands of this plant. The close-cropped fields were important breeding grounds for pintail (*Anas acuta*) and garganey (*A. querquedula*) and provided grazing for several species of geese, notably greylag (*Anser anser*), white-fronted (*A. albifrons*) and bean (*A. fabalis*), while the muddy margins, which were maintained by the puddling action of cattle hooves, attracted large numbers of wading birds. A further demonstration of this sort of trend in management has been made by Larsson (1969), who studied a series of grasslands around the shores of Swedish lakes. He made estimates of the qualitative changes in the breeding bird fauna which resulted from cessation of grazing and lowering of lake water level. He found that few species were common to both grazed and ungrazed grasslands and that the bird fauna was much more varied in the latter. Plate 18, taken from Larsson's paper, shows a lake shore at Dättern where the management of part of the marginal grassland has changed in recent years. In 1942 the whole area was cattle-grazed but by 1966 grazing had ceased on the northern part which had rapidly

Table 5.5 The distribution of some lake margin birds in Sweden in relation to grazing. Sites 1–20 are arranged in order of decreasing grazing pressure and increasing vegetation cover. X = species recorded (*after* Larsson, 1969)

Sites	1	2	3	4	5	6	7	8	9	10	11	12	13	14	15	16	17	18	19	20
Dunlin *Calidris alpina schinzii*	X	X	X	X	—	—	—	—	—	—	—	—	—	—	—	—	—	—	—	—
Skylark *Alauda arvensis*	X	X	X	X	—	X	—	—	—	X	—	—	—	—	—	—	—	—	—	—
Meadow Pipit *Anthus pratensis*	X	X	X	X	X	X	X	X	X	X	X	X	—	—	—	—	—	—	—	—
Blue-headed Wagtail *Montacilla f. flava*	X	X	X	X	X	—	X	X	X	X	—	X	—	—	—	—	—	—	—	—
Redshank *Tringa totanus*	—	X	X	—	—	—	—	—	X	X	—	—	—	—	—	—	—	—	—	—
Snipe *Gallinago gallinago*	—	X	X	X	X	—	X	X	X	X	—	X	X	—	X	—	X	—	X	—
Whinchat *Saxicola rubetra*	—	—	—	—	—	X	X	X	—	—	X	X	—	—	—	—	X	X	—	—
Willow Warbler *Phylloscopus trochilus*	—	—	—	—	—	—	—	—	—	—	X	—	X	X	X	—	X	X	X	X
Whitethroat *Sylvia communis*	—	—	—	—	—	—	—	—	—	—	X	X	—	X	X	—	X	X	—	—
Grasshopper Warbler *Locustella naevia*	—	—	—	—	—	—	—	—	—	—	—	X	—	—	X	—	X	—	—	—
Icterine Warbler *Hippolais icterina*	—	—	—	—	—	—	—	—	—	—	—	—	X	—	—	—	X	—	X	—
Chaffinch *Fringilla coelebs*	—	—	—	—	—	—	—	—	—	—	—	—	—	—	X	X	—	X	X	X

developed into a dense reed-bed. The difference in management is clearly shown in Plate 18.

In Table 5.5 12 species recorded by Larsson have been grouped by him in relation to their occurrence on sites subjected to different grazing pressures. In the areas influenced by grazing the number of species recorded was 23 and the corresponding figure for ungrazed areas was 38. The black-tailed godwit appeared to prefer areas with intensive to moderate grazing, while the dunlin (*Calidris alpina*) was found only in areas where grazing was still in progress. Larsson emphasized the importance, for some breeding birds, of the tussocky formation which often develops in wet grasslands grazed and trampled by cattle and said that pastures which are grazed evenly offer few nesting sites and have a poor bird fauna.

Cody (1968) made a detailed study of the distribution of birds in grassland communities in both North and South America and concluded that by using only two habitat indexes, vegetation height and its standard deviation, certain predictions could be made about the species present. Using data from North American grasslands he found he could predict (1) the number of species, (2) differences in feeding ecology and (3) their relative habitat separation in the community which occupies this habitat, for birdlife on South American grasslands. The predictions were found to hold regardless of species composition, grazing pressure, irrigation, age of grassland or latitude and Cody concluded that 'in each circumstance, selection appears to favour an ecological arrangement which is directly related to habitat structure in its simplest terms and which is affected by other variables only insomuch as they affect the vegetation height'.

Grazing Animals and their Associated Fauna

In this section we are concerned mainly with those animals which are directly dependent on grazing domestic animals, or rabbits where these are numerous enough to provide a continuous source of animal products which are exploited by other species. Morris (1969a) suggested the following as a habitat classification of such animal products:

    1. The living animal.
    2. Carrion (a) flesh
                 (b) bones
                 (c) fur and wool.
    3. Dung.
    4. Discarded fur and wool.

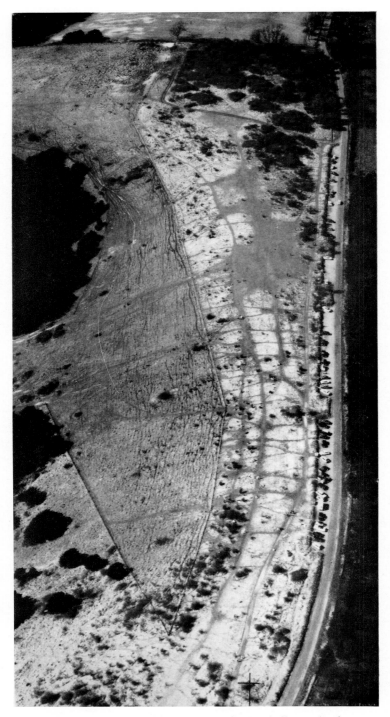

1. Old Winchester Hill NNR, Hants. A popular chalk grassland reserve where the many visitors have worn a network of paths across the vegetation. A car park extends along the roadside. (*Cambridge University Collection: copyright reserved.*)

2. The parkland at Upwood, Huntingdonshire. The ridge and furrow pattern was formed during the medieval period when

3. Old grassland in a 'sea of arable' at Lutton, Northamptonshire, shortly before it was destroyed by ploughing. (*Cambridge University Collection: copyright reserved.*)

4. Water-meadows at Breamore, Hampshire, on the river Avon. Photograph taken in January 1934. (*Photo J. Berry.*)

5. The same view in March 1970. The meadows have not been irrigated since 1949 and are currently used for rough grazing. (*Photo P. Wakely.*)

Pasque flower (*Anemone pulsatilla*) – a rare plant confined to chalk and oolitic grasslands
hich has decreased considerably as a result of habitat destruction. (*Photo M. C. F. Proctor.*)

7. Species-rich meadow grassland at Brampton, Huntingdonshire.

8. Flood meadow at Brampton, Huntingdonshire showing an abundance of great burnet (*Sanguisorba officinalis*). (*Photo P. Wakely.*)

. Fritillary (*Fritillaria meleagris*) – a characteristic plant of old flood meadows in southern England which are cut for hay. (*Photo P. Wakely.*)

(a) Small blue,
Cupido minimus

(b) Large blue,
Maculinea arion

10. Lycaenid butterflies of grasslands (a–d
(*Photos S. Beaufoy.*)

(c) Chalkhill blue,
Lysandra coridon

(d) Adonis blue,
Lysandra bellargus

11. Striped winged grasshopper, *Stenobothrus lineatus*, a common and characteristic species of grassland. (*Photo J. L. Mason.*)

12. Weeting Heath NNR, Norfolk. Breckland grass heath grazed by rabbits (within enclosure on right) contrasting with ungrazed area (on left). Note spread of *Pinus sylvestris* which began to colonise in 1954 after myxomatosis. (*Photo B. Forman.*)

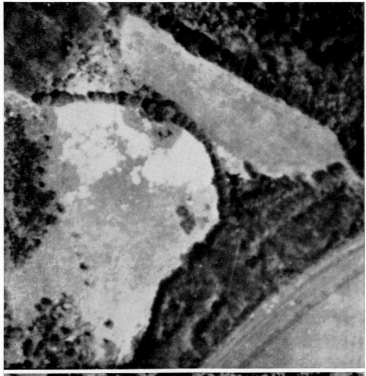

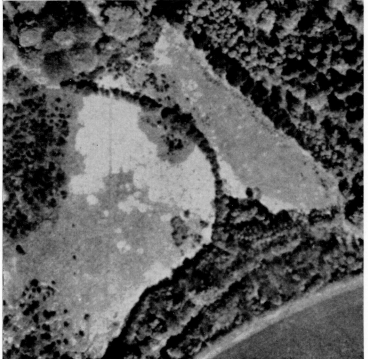

13–15. Wytham Woods, Berks. Spread of *Brachypodium pinnatum* (white areas) in relation to decline in rabbit grazing, 1946–1954.

13. Spring 1946. Rabbit population fairly low after war-time control measures. (*MOD Crown Copyright.*)

14. Spring 1953. Rabbit grazing moderate. The grass has clearly colonized new ground since 1946. (*Photo University of Oxford.*)

15. Spring 1960 after six years of negligible rabbit grazing. The population was virtually eliminated in 1954 by myxomatosis. (*Photo University of Oxford.*)

16. Redshank nest in wet meadow vegetation. (*Photo J. Robinson.*)

17. Grasshopper warbler at nest. (*Photo J. Robinson.*)

18. Lake margin at Dättern, Sweden. The southern half is cattle-grazed maintaining short grass and a muddy edge by the water. The northern half is no longer grazed and a dense reed-bed has formed, extending into shallow water. (After Larsson 1969. *Photo Geographical Survey Office of Sweden copyright reserved.*)

19. *Juniperus communis* (juniper), *Taxus baccata* (yew) and *Viburnum lantana* (wayfaring tree) on the chalk at Box Hill, Surrey. (*Photo P. Wakely.*)

20. Southern mixed shrubs with *Crataegus monogyna* (hawthorn) and *Rhamnus catharticus* (buckthorn) at Fetcham Downs, Surrey. (*Photo P. Wakely.*)

21. Blackbird (*Turdus merula*) feeding on hawthorn berries in winter. (*Photo Jane Burton.*)

22. A Cantharid beetle visiting hawthorn flowers. (*Photo J. A. Grant.*)

23. Autumn lady's-tresses (*Spiranthes spiralis*) – a geophytic orchid characteristic of grazed chalk grasslands. (*Photo P. Wakely.*)

24. Hill-fort at Hambledon Hill, Dorset. A site which combines high archaeological and biological interest. (*Cambridge University Collection, copyright reserved.*)

25. Cattle grazing at Port Meadow, Oxford. (*Photo P. Wakely.*)

26. Sheep grazing on a free range system on chalk grassland east of Chanctonbury Ring in Sussex. (*Photo P. Wakely.*)

27. Hay-making in the Fens. (*Photo B. Forman.*)

28. Mowing calcareous grassland with an Allen Scythe at Castor Hanglands NNR, Huntingdonshire. (*Photo P. Wakely.*)

29. Cutting limestone grassland with a 'Flymo', Castor Hanglands NNR. (*Photo P. Wakely.*)

30. Burning tor grass (*Brachypodium pinnatum*) grassland in February in order to destroy an accumulation of grass litter. Wye and

31. Scrub clearance by the British Trust for Conservation Volunteers at Wye and Crundale NNR. (*Photo H. M. Frawley.*)

32. Recording the cover of plant species in grassland vegetation using a point quadrat frame. (*Photo P. Wakely.*)

Various invertebrate animals are known to be feeders on these different animal products. The obligatory parasites of grazing animals can only be considered as grassland animals in a limited sense, but some of the parasitoids and facultative parasites, such as some species of *Lucilia* (blowflies), have more claim to be species of grassland animals because their free-living adult existence is spent there. Good farmers rarely allow their stock to die except in catastrophic circumstances and in these cases the carcasses are quickly destroyed. Dead rabbits, however, are of frequent occurrence, especially since the advent of myxomatosis, and rabbit carcasses usually support a varied fauna of Coleoptera and Diptera. The corpse of any sizeable mammal supports a considerable number of animals; Easton (1967) took 1967 beetles belonging to 87 species from a dead fox during the course of its decay. A few species were adventitious but most were either feeders on the corpse, or predators of the necrophagous beetles such as the Histeridae and *Aleochara* species. Easton's studies suggest that successional phenomena are marked in the ecology of the species feeding on dead animals. Such succession is well-known in the blowflies feeding on sheep carcasses. Although various insects are known to feed on bones and on fur and wool very little is known about these animals in the field.

Dung is in most pastures a more continuous resource than carrion. Both sheep and cattle dung support large faunas, predominantly of Coleoptera and Diptera. Little has been done on the range of species occurring in particular sites and in relation to such characteristics as succession. It is known that species of *Aphodius* are important in breaking down sheep dung, as these beetles do not occur in Australia, where the fouling of sheep pasture by their dung is an important problem. There are many species of *Aphodius*, some associated with the dung of particular mammals, and some occurring only in upland areas. The related Geotrupidae also feed as larvae on dung. In most cases burrows are excavated in grassland and dung rolled into these; eggs are then laid in the dung and the larvae develop underground. Cordylurid and other flies feed as larvae in dung and are attacked by beetles, both the predatory Histeridae (beetles) and semi-parasitic *Aleochara* spp. (Welch, 1965).

6 Ecological characteristics and classification of scrub communities

Scrub vegetation, or shrubland as it may be called in physiognomic-ecological classifications, is composed mainly of woody phanerophytes, both deciduous and evergreen (Ellenberg and Mueller-Dombois, 1966a and b). Ecologically the most interesting aspects of scrub are the dynamics of succession (Chapter 3) which lead, with time, to marked changes in the flora and fauna. Conservation of this habitat must take succession into account, and in this chapter an attempt is made to draw together some of the theory that is relevant to this overall aim. The biology of the woody species has been especially emphasized as they are so important in understanding the conservation ecology of scrub.

Classification of Scrub Communities

The delimitation of scrub communities from those of grassland and woodland is obviously arbitrary as they intergrade during seral changes. In this book, therefore, scrub vegetation is defined as extending from the stage at which the area of woody plants exceeds that covered by grassland, to that when woody plants reach 7 m in height and are composed mainly of tree species.

A detailed classification of scrub types in Britain cannot be made yet because of the lack of knowledge of the range of variation in this vegetation. However, it is possible to recognize some of the broader divisions that are associated with base content, soil, water, and geographical location which relate to climatic factors. Further variation is superimposed by the earlier management of the grasslands and the point at which management of the grassland ceases and seral progression towards woodland begins. Even in the case of apparently similar conditions there can be several possible

124

dominants whose similar component species can be segregated in pure stands or in mixtures. For example, the dominant species *Prunus spinosa* and *Ligustrum vulgare* replace one another without affecting the associated species on maritime cliff tops in west Cornwall (Malloch, 1971). Scrub communities may also have originated from areas that are not primarily grasslands such as bare chalk in quarries (Hope-Simpson, 1940) and abandoned ploughed fields (Lloyd and Pigott, 1967). In these cases the seres may not pass through a grassland phase but be composed mainly of dicotyledons and woody species. Fragmentary scrub communities, many of them subject to strong edge effects and frequent disturbance, are also familiar in the agricultural landscape in hedges, road verges, railway embankments and woodland edges. Hedges are normally planted and then managed, but over a period of time there appears to be a trend towards invasion by a greater variety of woody species (Hooper, 1970). The scrub communities related to management by coppicing are not considered here as they belong more to the woodland ecosystem.

The broader groups of scrub types can be assigned to the higher divisions of the phytosociological systems used on the continent. Thus scrub formations on calcareous soils usually belong to the class Querco–Fagetea Br.–Bl. et Vlieg. 37, order Prunetalia Tx. 52, alliance Berberidion Br.–Bl. 50. Shimwell (1971a, b) has described some of the scrub communities in this order which are seral from the grasslands of the order Brometalia erecti Br.–Bl. 36 on limestones in Britain. Scrub communities on dry neutral soils also belong to the order Prunetalia Tx. 52. On the wetter neutral soils they belong mainly to the classes Salicetea purpurea Moore 58, and the Alnetea glutinosae Br.–Bl. et Tx. 43. Scrub formations on the dry acid soils fall into the class Nardo–Calluneatea Prsg. 49 and the Erico–Pinetea Horv. 59.

An overall picture of the most likely course of succession from the main grassland groups through scrub to woodland is illustrated in Fig. 6.1. This shows the important species of woody plants that are found in scrub communities in southern and eastern England on the main soil types. The common dominants are shown for each type, and any one of these species may predominate on particular sites or parts of sites. There may also be variants that cannot be placed easily into any one of the seven types shown. Shrubs may be mixed with varying proportions of tree species, and this influences the later succession which may pass almost directly to woodland, or else it may pass through a longer phase of dominance by shrubs. The diagram in Fig. 6.1 is simplified to show selected tree species as the climax woodland species for each soil type, although in reality there are

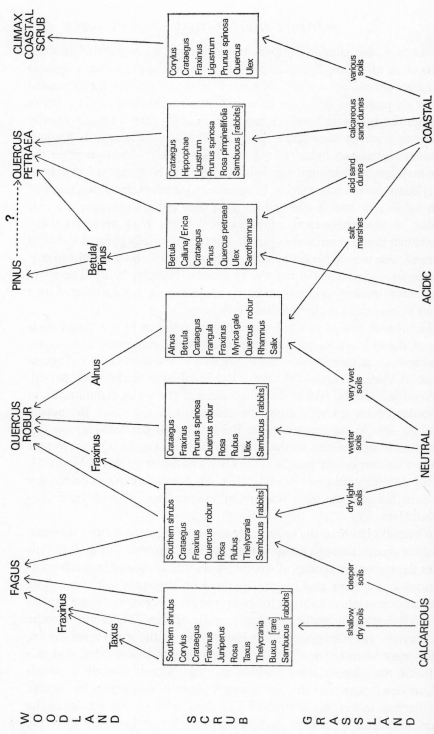

Fig. 6.1 Diagram to show the most likely course of succession from the main grassland types through scrub to woodland in southern and eastern England.

likely to be a range of mixtures of these dominant species. The exact relationships of these trees to each other under natural conditions is not understood. For example it is uncertain whether *Pinus sylvestris* or *Quercus petraea* would predominate on light, dry, sandy soils in the south of England. Nearly all unmanaged scrub communities do eventually pass to woodland, only where severe exposure retards the growth of the trees does a climax community occur. In the southern lowlands this is only in maritime areas where climax scrub is found on exposed cliffs. Elsewhere climax scrub occurs beyond the northern and altitudinal limits for tree survival. Inland rocky areas may also support a type of scrub climax.

Nearly all scrub succession in lowland Britain is secondary (Chapter 3) although there may be fragmentary areas on coastal sites and cliffs and in the freshwater hydroseres where succession is primary and man has had no influence. Primary succession may be seen in some areas where *Hippophaë rhamnoides* is invading accreting sand dunes and increasing the suitability of the soil for other colonizing species by its ability to fix nitrogen and bind the sand. Another pioneer woody species, *Myrica gale*, shows this nitrogen-fixing ability in hydrosere successions. Primary succession is often extremely slow in progressing to a climax, and this can be seen in cliff vegetation such as that described for Magnesian limestone grips by Jackson and Sheldon (1949). Here, cliff recession is caused by *Taxus* and other woody species and the rock communities can be regarded as climax vegetation although eventually, and together with other forces of erosion, the physiography of the area will be changed.

In the classification which follows, an attempt has been made to describe the scrub communities according to dominant species or mixtures of species of shrubs and trees. These have been listed under their most characteristic soil type. Only the main shrub types have been described. The woodland tree species have been included as part of the variation of these communities, and have been considered in the context of the most likely course of succession of the communities towards woodland. Scrub communities that are entirely dominated by woodland tree species have not been described. The succession from acidic grasslands to dwarf-shrub heaths and their later stages of progression through scrub towards woodland are not considered in this book; they are described in some detail in Gimingham (1972). Coastal scrub is also considered only very briefly. More information is available for southern scrub which has been studied in greater detail.

Calcareous soils

Corylus avellana (Hazel)

Corylus avellana is commoner in the north and west and is particularly associated with limestone. In Derbyshire it is an important species in various scrub communities, especially the 'Retrogressive scrub' of Moss (1913) where there is a very rich field layer including *Melica nutans*, *Convallaria majalis* and *Geranium sanguineum*. Succession is frequently to *Fraxinus*, although occasionally *Acer pseudoplatanus* may be present (Merton, 1970). In northern and western Scotland *Corylus* occurs with *Betula pubescens*, *Fraxinus excelsior*, *Quercus* spp., *Prunus padus*, and *Sorbus aucuparia*. Many of these communities are montane, and have been described by McVean and Ratcliffe (1962). *Corylus* is also often abundant in maritime areas, and in the Burren of Ireland forms a climax community in the *Corylus avellana–Oxalis acetosella* association (Ivimey-Cook and Proctor, 1966).

On neutral, but not water-logged, soils *Corylus* is often an important species in oakwoods, as on the clay in the Midlands and East Anglia. It may have been planted in many of these woodland situations, but it does very occasionally invade grasslands in these areas although far less commonly than *Crataegus monogyna* or *Quercus robur*.

Juniperus communis (Juniper)

Juniperus occurs as a dominant or co-dominant in communities on limestone soils throughout Britain, although these communities are more frequently found in southern England. The commonest woody species that grows with juniper is *Taxus baccata* (Plate 19), but many other species of southern shrubs may be present. The field layer is usually dominated by *Festuca ovina/rubra* or *Zerna erecta*. One variant of the juniper association occurs as very old juniper bushes mixed with *Sambucus nigra*, and is probably related to heavy rabbit grazing in the past. Succession is frequently to *Taxus baccata*, although occasionally to *Fraxinus* or *Fagus*. Details have been described for the yew communities of the South Downs by Watt (1926), and the Chiltern beechwoods (Watt, 1934). A very detailed survey of the present status of juniper in southern England has been carried out by Ward (1973), and Fig. 6.2 shows the populations at present found in Surrey, together with extinct sites.

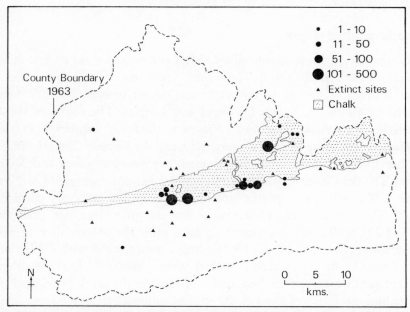

Fig. 6.2 Juniper in Surrey showing populations found in each square kilometre in 1968–71, and all records of extinct sites.

Juniperus also occurs in heathlands, with *Pteridium*, and in acid grass-lands in northern England and Scotland. It is now so rare in these communities in the south of England that a true scrub is not formed. These communities and their variations are further discussed in Gimingham (1972).

Taxus baccata (Yew)

Taxus occurs on calcareous soils in southern Britain, where it is often closely associated with the southern juniper scrub, although it may also be invasive with *Crataegus monogyna* and *Fraxinus* (Watt, 1926). It also occurs in the absence of juniper, on rocky areas and screes as on the Carboniferous limestone of the Mendips and Derbyshire. In the later stages of succession the dense shade of the closed yew canopy suppresses virtually all the field-layer species. *Taxus* communities are thought to be seral to *Fraxinus*, but the species is very long-lived and may persist as woodland for hundreds of years. The yew woodland at Kingley Vale, Sussex, is one of the best examples in Europe of a yew wood at different stages in the succession. In the Lake District *Taxus* communities are found on soils which are only moderately base-rich.

Southern mixed shrubs

The species of the southern mixed shrub communities are typical of the calcareous soils in the south. These shrubs may be the dominant species of stands of scrub, or may be important constituent species of stands dominated by *Juniperus, Taxus, Crataegus* and *Corylus*. The southern shrubs include *Thelycrania sanguinea, Viburnum lantana, Ligustrum vulgare, Rhamnus catharticus, Euonymus europaeus,* and the climber *Clematis vitalba.* Other species that are frequently present are *Rosa rubiginosa* and *Sorbus aria.* The field layer is usually dominated by *Festuca ovina/rubra* or *Zerna erecta.* The relative proportions of the shrub species may vary considerably. On deeper soils, *Crataegus monogyna* often becomes dominant (Plate 20), while on certain nutrient-poor soils *Thelycrania* may become dominant (Lloyd and Pigott, 1967). Similar communities with *Thelycrania* are found on limestone screes in Derbyshire (Shimwell, 1968). Southern mixed shrub communities become less diverse in the north because of the distributional limits of some of the species, although there are some interesting sites on limestone around Morecambe Bay. *Viburnum lantana* only extends to Derbyshire, *Thelycrania* and *Rhamnus* reach Cumberland and Durham, while *Euonymus* reaches its limits in southern Scotland. Succession in these communities is most frequently to *Fraxinus*, although it may also be to *Taxus, Quercus* or *Fagus.*

Buxus sempervirens (Box)

This species forms an unusual type of scrub on the calcareous soils in the south, where it may be dominant or associated with *Sambucus, Taxus* or *Fraxinus.* It is found on very steep slopes on two of its best-known sites, and it has been suggested that at Box Hill, Surrey, in areas where the trees do not form a closed canopy, open-ground species could have persisted when the rest of the country was forest-covered (Pigott and Walters, 1954). Elsewhere the trees form a very dense shade and few field-layer species survive, although *Mercurialis perennis* is often present. Recent opinion favours the box as being a native species in Britain (Pigott and Walters, 1953; Staples, 1970).

Local species of *Sorbus* (Whitebeam)

Local species of *Sorbus* never become dominant stands in scrub, but usually grow mixed with other calciphilous species of shrubs and trees on rocky limestone outcrops in western and northern Britain. Some species are endemic and are exceedingly rare. For example *S. minima* and *S. leptophylla* are found only in Breconshire, and *S. bristoliensis* is found only in the Avon Gorge. Other species such as *S. rupicola* are more widely dispersed. On many of these rocky cliffs true forest canopy is unlikely ever to have developed, and the cliff vegetation can be regarded as a climax.

Sorbus aria is included under the southern mixed shrub communities where it is a common species.

Sambucus nigra (Elder)

Sambucus scrub is strongly associated with heavy rabbit grazing on most soils and it is especially characteristic of the chalk soils around rabbit warrens. *Sambucus* also marks the site of badger setts. This pattern of distribution is due to its preference for soils with a high nutrient concentration and for disturbed ground which favours the germination and growth of the seedlings. It is unpalatable to both rabbits and sheep and this is a further reason for its association with rabbit warrens on grazed downland. *Sambucus* is frequently found in nearly pure stands, or it may form rings around other older trees, the latter being related to the dispersal of its seeds by birds. The typical field-layer species include *Poa trivialis*, *Mercurialis perennis*, *Chamaenerion angustifolium* and *Urtica dioica*. Epiphytic mosses are also important locally. Succession in these communities, at least on the calcareous soils, seems to be towards *Fraxinus*.

Neutral soils

Crataegus monogyna (Hawthorn)

Crataegus monogyna is the commonest scrub species in lowland Britain. It occurs on many different soils, but avoids the more acid, very dry or very wet soils. On the heavier soils, such as on boulder clays, *Crataegus* is usually the dominant species on abandoned fields. It is associated with species of *Rosa* and *Rubus*, and with *Prunus spinosa*, *Ulex europaeus*, *Quercus* spp., and *Fraxinus* in successions towards oak woodland. If sufficient oaks invade into the grasslands at the same time as the *Crataegus* the succession may pass directly to woodland, otherwise there may be a longer period of dominance by scrub. On the calcareous or neutral soils with impeded drainage *Q. robur* may become the more abundant species, while on drier and more acid soils it is replaced by *Q. petraea*. In the later seral stages of these scrub communities, *Corylus*, *Acer campestre* and *Carpinus* may also play a part. Some of these variations in the sere leading to oak woodland in Essex have been described by Adamson (1932).

The communities of southern shrubs in which *Crataegus* may be a dominant or co-dominant species on calcareous soils have been described above. Those communities on acid soils with *Ulex* are described under that species.

Prunus spinosa (Blackthorn)

Scrub with *Prunus spinosa* is associated especially with neutral to acid soils on clay, in and adjoining woods. *P. spinosa* is frequently mixed with scrub of *Crataegus monogyna* and the communities intergrade, with *Prunus* being the more successful in slightly wetter areas than *Crataegus*. Its habit of suckering is important in spreading from woodland, and it is much less invasive into abandoned grasslands than *Crataegus*. Succession tends to be to *Quercus robur*.

Salix spp. and *Alnus glutinosa* (Willows and alder)

Succession from grasslands which have a high water table has been little studied and several different communities could be defined, although only one overall type is considered here. The most important invasive woody species are *Alnus glutinosa*, *Salix cinerea* and other species of *Salix*, and *Myrica gale*; while *Frangula alnus*, *Viburnum opulus*, *Rhamnus catharticus*, *Betula pubescens* and *Fraxinus excelsior* may frequently be present. Field-layer species in the carr communities include *Cladium mariscus*, *Dryopteris thelypteris*, *Carex paniculata*, *C. acutiformis*, *Chamaenerion angustifolium*, *Urtica dioica*, and *Phragmites communis*. Where valley fen communities have been converted to grasslands and then abandoned, as in the Breck fens, the growth of tall species such as *Chamaenerion* and *Urtica* may be marked, and it becomes difficult for woody species to invade. Succession to *Salix* and *Alnus* is quicker where there has been less interference (Haslam, 1965). *Alnus* is less adapted to the invasion of grasslands, unless they have been so neglected that drainage is interrupted, and there is seasonal wetness (McVean, 1956). *Alnus* does not usually succeed an earlier *Salix* or *Rhamnus/ Frangula* carr, but may invade at the same time, and as it grows taller and lives longer it eventually becomes dominant. The later succession of these carrs is not fully understood. They apparently pass to climax mixed wood-land with *Quercus robur* and *Betula pubescens* on drier areas, or in wetter areas the alder carr may remain as a climax. In very wet situations there may be cyclic changes, with the alderwood passing back to marsh eventually (McVean, 1956).

Acid soils

Ulex spp. (Gorse)

Communities dominated by species of *Ulex* are common throughout the country: *U. europaeus* is generally widespread, *U. gallii* occurs mainly in the south and west, while *U. minor* is a constituent of dwarf-shrub communities in the south and east. Stands of *Ulex* are often associated with heathland (Gimingham, 1972) which is outside the scope of this book, but they are also found in successions from acid grasslands and from chalk and other limestone heaths. In the more typical case of succession on acid grasslands they are often extreme variants of the succession to oak woodland already described for *Crataegus*, these communities intergrading with each other. The same range of species is found, including *Prunus spinosa*, *Rubus idaeus* and other *Rubus* and *Rosa* species, and the two species of *Quercus*. In drier situations *Sarothamnus* may also be present, and the trees, *Betula* spp. and *Pinus sylvestris*, may be important, the former especially where there has been burning. Eventual succession may be to *Pinus* in these cases. *Ulex* and *Sarothamnus* are often associated with disturbed land, and in the New Forest stands of *Ulex* have been found to mark the limits of old cultivations (Tubbs and Jones, 1964).

Calcareous heath scrub is not very common; it occurs on soils derived from loessic material, from the edges of clay-with-flints deposits, or from coombe deposits (Grubb, Green and Merrifield, 1969). Characteristic communities have both acidophilous species such as *Ulex* and *Calluna* together with calciphilous field-layer species, and shrubs such as *Viburnum lantana*, *Rhamnus catharticus* and *Ligustrum vulgare*. In some instances *Ulex* is present but *Calluna* is absent; several such stands can be seen on the chalk of the Isle of Wight. Limestone heaths with *U. gallii* occur in the west of Britain.

Coastal scrub

Cliff-top scrub

This type describes a range of communities on the different soils and corresponds to many already described, but with the additional factor of maritime exposure. *Ulex* spp., *Corylus avellana*, *Prunus spinosa*, *Ligustrum vulgare*, *Rosa* and *Rubus* are particularly common, and some cliff-top scrub communities in Cornwall have been studied by Malloch (1971). On lime-stone cliffs *Juniperus communis* may be present as well as *Rosa pimpinelli-folia*, and the introduced but naturally spreading shrub *Cotoneaster microphyllum*. The native *C. integerrimus* occurs only on one site in North Wales. Succession does not usually proceed to woodland because of exposure factors, and the shrubs and trees are often severely windpruned.

Sand dune scrub

This type also describes a number of different communities depending on the soil water and base content, and presumably also on the sand dune com-munity that is invaded by the woody plants, but relatively little seems to be known about the invasion of dune grasslands. Communities with *Hippophaë rhamnoides*, the sea buckthorn, appear to be native on the east coast, but since the loss of the rabbits by myxomatosis the plant has successfully spread in many dune systems in other parts of the country including areas where it was formerly planted as a sand-stabilizer (Ranwell, 1972a). *Hippo-phaë* is associated with *Salix repens, Sambucus nigra, Prunus spinosa, Ligu-strum vulgare, Rosa pimpinellifolia*, and other *Rosa* and *Rubus* species. Some of these communities with woody species have been described for dunes in Holland by Westhoff (1952) and by van der Maarel and Westhoff (1964). On the more acid sand dunes *Hippophaë* is less common, and succession through heathland and *Ulex* may occur. The later succession towards woodlands on sand dune systems is very poorly understood in Britain, because nearly all areas with fixed dunes have been used by man for plan-tations or have been converted to dune grasslands.

Shingle scrub

Scrub on shingle often suffers from very severe maritime exposure, and there is dwarfing and windpruning of the bushes. *Sarothamnus scoparius* var *prostratus*, together with *Solanum dulcamara* var *marinum, Prunus spinosa* and *Suaeda fruticosa* occur together with *Ulex europaeus, Sambucus* and *Rubus* spp. in more sheltered situations. The introduced shrub *Tamarix gallica* has also spread in some areas. On the well-known shingle site of Dungeness, there are areas with the prostrate broom, and also a unique holly wood which has been described by Peterken and Hubbard (1972). Succession from shingle scrub towards woodland is not under-stood, as again there are only fragmentary remains of seres.

Establishment of Scrub Vegetation

The presence of scrub is closely related to the patterns of land use and management of grasslands and other areas in the past. The most critical phase is that of establishment, for once the woody plants are present and have made a certain amount of growth they can often withstand grazing pressure. Scrub stands, therefore, can be dated by historical events when there were changes in the management of the land. The most important of these changes in recent years has been that of the loss of rabbits by myxomatosis, so that in 1955 the former intense grazing pressure was relaxed (Chapter 5). Woody plants were able to invade many areas, including especially dry situations on chalk downs, heathlands and sand dunes, which had been maintained previously as plagioclimax communities by the rabbit grazing. In many of these areas shrubs and trees have continued to grow unchecked, and on some areas, such as the Derbyshire limestone dales (Merton, 1970), are rapidly becoming woodlands. In the case of other scrub stands, farming economics may have been an important factor contributing to their establishment. For example the even-aged *Crataegus* on some upland *Festuca-Agrostis* grasslands appears to be related to a period when sheep were reduced in numbers on the hill farms following a fall in wool and meat prices. Similarly, fragmentary remains of old *Crataegus* scrub can be found on the heavy clay of the Midlands, and these date back to the inter-war period when it was uneconomic to cultivate these soils. Another example is the regeneration of birch and rowan in north-west Scotland which can be related to a reduction of grazing pressure in marginal areas due to the present decline in the crofter population (Bunce, 1970).

In the early stages of invasion by shrubs, the dispersal mechanisms of the seeds are obviously very important. Many woody species have brightly coloured fleshy fruits adapted for dispersal by birds, and in autumn they provide food for blackbirds, song and missel thrushes as well as flocks of migrant fieldfares and redwings (Plate 21). These species show some preferences in their diets, but this is modified according to the availability of the different fruits (Hartley, 1954). Rodents are also involved in seed dispersal due to their habit of forgetting 'stores' of seeds, but they also cause severe mortality of the seeds, destroying very high numbers of hazel nuts, for example (Sanderson, 1958), so that the difference between dispersal agents and predators on seeds cannot always be distinguished. The balance between the production of seeds by individual plants and the numbers of predators on them may be important in influencing the rate and direction of plant succession (Janzen, 1971).

A common observation is that the influence of seed parents is strong locally, so that, where one species outnumbers all the others, it has a greater chance of establishing under suitable conditions. Local imbalance of seed parents can be very great, and an outstanding example is the vast seed resource of *Crataegus* in hedges, a factor which probably accounts for the preponderance of this species in scrub. On the other hand, *Juniperus* has become so rare in some districts that it is unlikely to be important in scrub communities however suitable are the conditions for germination and later growth of the young plants. Godwin (1936) studied the relationship between *Frangula alnus* and *Rhamnus catharticus* at Wicken Fen and came to the conclusion that although he could demonstrate differences in the

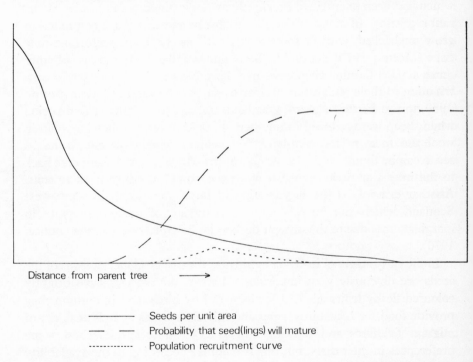

Distance from parent tree ⟶

—————— Seeds per unit area
— — — Probability that seed(lings) will mature
············· Population recruitment curve

Fig. 6.3 Model diagram from Janzen (1970) to show that with increasing distance from the parent tree the number of seeds per unit area (——————) declines, but the probability that a dispersed seed or seedlings will be missed by the host-specific seed and seedling predators before maturing increases (- - - - -). The product of the I and P curves yields a population recruitment curve (......) with a peak at the distance from the parent where a new adult is most likely to appear.

removal of the berries of the two species by birds, the overriding influence was the great number of seed parents of *Frangula*. Forgotten 'stores' of seeds by mice were found to have helped in the dispersal of both species.

The distances over which the local influences of seed parents extend have hardly been studied in this country, but there is some interesting work on South American woody species by Janzen (1970). He considered the relationships between the numbers of seeds produced from a parent tree and the distance to which the seeds were likely to be dispersed, that is the 'seed shadow', which fell off with distance. This was compared to the activity of the seed and seedling predators which tended to be more important near to the parent trees where the numbers of seeds were greater. Thus the probability that a seedling would mature was greater at a distance from the parent. In fact there was an area where the seeds were more likely to mature according to the balance of these various factors (Fig. 6.3). The seed shadow could be distorted by local conditions of topography which influenced the predators and dispersal agents. In the case of seeds that are dispersed by wind the seed shadow is often closer to the parent plants (Janzen, 1971), and this can be observed in this country for *Pinus* and *Fraxinus*. Heavier seeds tend to be dispersed by mammals, and Ridley (1930) thought that these seeds were not likely to be transported for great distances. He found that squirrels carried acorns, hazel nuts and beech mast for up to about 30 m. Mellanby (1968) found that oak regeneration levels had not fallen off at 200 m from the parent trees in a wood. Over longer distances bird dispersal seems to be the most important factor and oceanic islands over 100 miles (160 km) from the mainland may have bird-dispersed plant species growing on them.

Scrub establishment is influenced by the previous management of the area and its condition at the time when grazing or other restraint is removed. Two factors are involved: first, the presence of woody plants already in the grassland, and secondly those species that are able to establish once management ceases. Where grazing has been continuous a selection pressure has been operating on any seedlings or young plants, and those that are able to survive the damage caused by grazing persist in the sward as short, coppiced plants. *Crataegus* plants only a few centimetres high have been found to be growing very slowly, and to be up to 20 years old. When grazing pressure is reduced these plants are able to grow unchecked. Most shrubs do, in fact, show evidence of adaptations to resisting grazing, some having an armoury of spines or thorns (Fig. 6.4), others such as *Prunus spinosa* and *Thelycrania* showing suckering and a good ability to regrow from the base, while others are unpalatable, for example *Sambucus*.

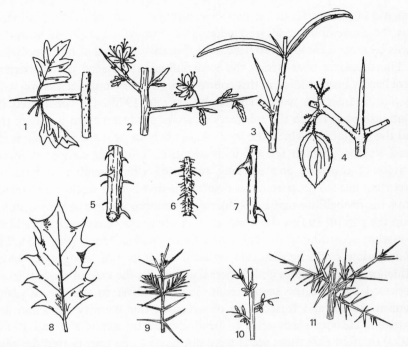

Fig. 6.4 Thorns and spines of woody plants of scrub:
1. *Crataegus monogyna* (hawthorn); 2. *Prunus spinosa* (blackthorn); 3. *Hippophaë rhamnoides* (sea buckthorn); 4. *Rhamnus catharticus* (purging buckthorn); 5 *Rubus fruticosus* (blackberry); 6. *Rosa pimpinellifolia* (burnet rose); 7. *Rosa canina* (dog rose); 8. *Ilex aquifolium* (holly); 9. *Juniperus communis* (juniper); 10. *Genista anglica* (needle furze); 11. *Ulex europaeus* (gorse).

The last species has a strong selective advantage in the severely grazed conditions around rabbit warrens. There are also interactions between the various woody species under grazed conditions, and the more susceptible species may be protected by growing in the middle of other species. Both *Juniperus communis* and *Ulex europaeus* are thought to act as 'nurses' to more palatable species like *Crataegus monogyna* (Tansley, 1953). Once these more susceptible species reach a certain height out of reach of the animals, subsequent grazing does not check them. They may grow on and even eliminate the original *Juniperus* or *Ulex* by shading them.

Once management ceases the condition of the grassland or other habitat has a considerable influence on the subsequent successful colonization by different species. Shrubs are often independent initially of the nutrient status of the soil because of the greater storage of food in their larger seeds, but successful later establishment is affected by the nutrient status of the soil and by the cover and vigour of the other vegetation (Lloyd and Pigott 1967). *Juniperus*, for example, appears to need a habitat with little competition from other plants and has been found colonizing in various open ground situations, such as abandoned ploughed fields, old quarries and banks on newly made roads (Vedel, 1961; Fitter, 1968; Ward, 1973). It is able to survive better than *Crataegus* on drier, shallower and nutrient-poor soils on the chalk (Watt, 1934). Where succession begins from arable areas the field layers may include arable weeds for long periods of time (Salisbury, 1918), and it has already been noted that there are likely to be fewer grasses in these entirely ungrazed situations.

Changes in Plants during Succession

As the sere progresses from grassland the shrubs and trees begin to grow and competition reduces the numbers of low-growing species and those preferring open ground, while at the same time, and especially on deeper richer soils, taller species begin to invade. As the shrubs and trees continue to grow they begin to shade the ground so that in places many of the former grassland species die out. Bare ground may appear and is then often colonized by shade-tolerant woodland species. The increasing structural complexity results eventually in tree, shrub, field and ground layers. As an example, Table 6.1 shows the species recorded in twelve stands of scrub of *Crataegus monogyna* on calcareous soils in southern England. A 10 m × 10 m quadrat was used, and the woody species assessed on a Domin scale; the field-layer species were recorded on a presence/absence basis, and species growing within another 10 m of the quadrat were also noted. The stands are divided into three groups depending on whether the canopy of the bushes in the quadrat was predominantly open, patchy or closed. An attempt has also been made to categorize the field-layer species into those typical of shorter grassland, marginal or 'edge' situations and woodland. The marginal species are found in transition zones or ecotones between grasslands and woodlands (Chapter 3), and many of them are common in hedgerows and other fragmentary scrub areas. The changes in the field-layer species are quite marked, and there is a general decline in the numbers as the sere progresses. In the early stages the proportion of grassland species is quite high, and while the marginal species are found in all stages they are commoner in the intermediate patchy stage. The few woodland species are all found in the closed scrub, and they are present only in low numbers. An explanation for this may be that many woodland species are very slow colonizers. Rackham (1971) showed that there were relatively few

Table 6.1 Species of plants recorded in 10 × 10 m quadrats of open, patchy and closed canopy scrub of *Crataegus monogyna* on calcareous soils in southern England (Woody species on the Domin scale. 0 = field layer species present in quadrat, † = species nearby, within 10 m of the quadrat, X = woody species on Domin scale, 9–X see Table 12.2).

	Open				Patchy				Closed			
Sample No.	1	2	3	4	5	6	7	8	9	10	11	12
Grid reference	SP	SP	TQ	TQ	TQ	TQ	TQ	SZ	SP	TQ	TQ	SU
	961157	964171	318134	305137	102485	324133	332537	593868	961156	505012	380124	937225
Aspect (°)	105	350	0	0	180	270	220	170	120	280	315	120
Slope (°)	0	5	15	15	20	15	30	10	0	35	30	10
Soil depth (cm)	11·4	22·9	7·6	5·1	22·9	15·2	20·3	7·6	>30·5	>30·5	25·4	10
pH	—	7·85	7·2	7·2	7·8	7·7	7·7	7·2	—	7·1	7·3	—
Height (m)	2	2	2	1·5	1-2	1·5	1-2	1	4-5	7	3·4	5
Management	None	None	Grazed	Grazed	Grazed	None	None	None	None	None	None	None
Acer pseudoplatanus	—	—	—	—	—	—	—	—	—	—	—	—
Clematis vitalba	—	—	†	—	x	†	†	†	—	—	1	—
Corylus avellana	5	7	4	2	—	—	—	—	—	—	2	8
Crataegus monogyna	—	4	4	1	6	5	5	5	9	9	9	8
Fraxinus excelsior	—	†	†	1	—	3	†	†	1	—	†	†
Hedera helix	—	—	—	—	—	—	—	—	—	1	†	—
Ligustrum vulgare	—	†	†	—	2	—	—	—	—	x	x	1
Lonicera periclymenum	—	—	—	—	—	—	—	—	—	—	—	†
Prunus avium	—	x	—	—	—	—	—	—	—	—	—	—
P. spinosa	x	—	—	—	—	†	—	†	2	†	—	8
Quercus robur	—	—	†	x	—	—	—	†	†	†	†	†
Rhamnus catharticus	—	†	x	—	5	2	2	2	—	—	—	—
Rosa canina	—	1	—	—	5	2	2	4	—	—	—	—
R. rubiginosa	—	—	†	†	5	—	†	†	—	—	†	—
Rubus fruticosus	—	†	4	4	5	—	†	†	†	†	†	—
R. idaeus	—	—	—	—	—	—	—	—	7	—	†	—
Sambucus nigra	—	†	†	x	—	—	—	—	7	4	x	—
Sorbus aria	—	x	—	x	—	x	—	—	—	—	—	—
S. aucuparia	—	—	—	—	—	—	—	—	—	—	—	—
Tamus communis	—	—	—	—	x	x	—	—	—	—	x	—
Taxus baccata	—	†	†	†	—	—	—	†	—	—	—	†
Thelycrania sanguinea	3	—	†	x	2	x	—	†	—	—	—	—
Ulex europaeus	—	—	—	—	—	—	—	—	—	—	—	—

Field-layer species

	Species
M	*Arrhenatherum elatius*
	Brachypodium pinnatum
M	*B. sylvaticum*
G	*Briza media*
G	*Cynosurus cristatus*
	Dactylis glomerata
	Deschampsia cespitosa
G	*Festuca arundinacea*
G	*F. ovina*
G	*F. rubra*
G	*Helictotrichon pratense*
G	*H. pubescens*
G	*Holcus lanatus*
G	*Koeleria gracilis*
G	*Phleum bertolonii*
G	*Poa pratensis*
W	*P. trivialis*
M	*Trisetum flavescens*
G	*Zerna erecta*
G	*Carex flacca*
W	*Dryopteris filix-mas*
M	*Achillea millefolium*
M	*Agrimonia eupatoria*
W	*Arcticum minus*
W	*Arum maculatum*
G	*Campanula rotundifolia*
M	*Centaurea nigra*
G	*Centaurium erythraea*
G	*Cerastium arvense*
M	*Chamaenerion angustifolium*
	Chrysanthemum leucanthemum

141

Table 6.1 *continued*

	Sample No.	Open				Patchy				Closed			
		1	2	3	4	5	6	7	8	9	10	11	12
G	*Cirsium acaulon*	–	0	0	–	–	–	–	–	–	–	–	0
	C. arvense	+	–	–	–	–	–	–	+	–	–	–	–
M	*C. vulgare*	–	–	–	–	0	–	0	–	–	–	–	–
M	*Clinopodium vulgare*	–	–	0	–	0	–	0	–	–	–	–	–
G	*Crepis capillaris*	–	0	–	–	0	–	–	–	–	0	–	–
G	*Dactylorchis fuchsii*	–	–	0	0	–	–	–	+	–	–	–	–
	Daucus carota	–	–	–	–	–	–	–	–	–	–	–	–
M	*Epilobium montanum*	–	–	–	–	–	–	–	–	–	0	–	–
M	*Eupatorium cannabinum*	–	–	–	0	–	–	–	–	–	–	–	–
G	*Euphrasia nemorosa*	–	–	0	0	–	–	–	–	–	–	–	–
G	*Filipendula vulgaris*	0	–	–	–	–	–	–	–	–	–	–	–
M	*Fragaria vesca*	–	–	0	0	–	–	–	0	–	–	–	0
M	*Galium aparine*	0	0	0	0	–	–	0	–	0	–	–	0
G	*G. mollugo*	0	0	0	0	0	–	0	0	–	0	–	–
G	*G. verum*	–	–	–	–	–	–	0	–	–	0	–	0
G	*Gentianella amarella*	–	–	–	–	–	–	–	–	–	–	–	–
W	*Geranium robertianum*	–	–	–	–	–	–	–	–	–	–	–	–
W	*Geum urbanum*	–	–	–	–	–	–	–	–	–	–	–	–
W	*Glechoma hederacea*	–	0	0	0	–	–	–	0	–	–	–	0
G	*Gymnadenia conopsea*	–	–	–	–	–	–	–	–	0	0	–	0
W	*Hedera helix*	0	0	0	0	0	0	0	0	–	0	–	0
G	*Helianthemum chamaecistus*	+	–	–	–	–	–	–	–	–	–	–	–
M	*Heracleum sphondylium*	0	0	0	0	0	0	0	–	–	0	–	0
G	*Hieraceum* sp.	–	0	0	0	0	–	0	–	–	–	–	–
G	*Hippocrepis comosa*	–	–	–	–	–	–	–	–	–	–	–	–
M	*Hypericum perforatum*	–	–	–	–	–	–	–	–	–	–	–	–
M	*Knautia arvensis*	–	–	0	0	0	–	–	+	–	–	–	0
G	*Leontodon hispidus*	–	0	0	0	0	–	–	–	–	–	–	–
G	*Linum catharticum*	–	–	0	–	0	–	–	–	–	–	–	–

142

		19	32	42	31	30	31	35	13	10	12	6	17
G	*Lotus corniculatus*	o	o	o	o	o	o	o	+	–	–	–	–
G	*Medicago lupulina*	–	–	o	–	o	+	o	–	–	–	–	–
M	*Ononis repens*	–	–	–	–	–	–	–	–	–	–	–	–
M	*Origanum vulgare*	–	–	–	–	–	–	o	–	–	–	–	–
M	*Pastinaca sativa*	–	–	–	–	–	–	o	–	–	–	–	–
G	*Phyteuma tenerum*	–	–	o	o	–	–	o	–	–	–	–	–
G	*Pimpinella saxifraga*	–	o	o	o	–	o	–	o	–	–	–	–
G	*Plantago lanceolata*	o	o	o	o	–	o	–	–	–	–	–	o
G	*P. media*	–	o	–	–	o	–	o	o	–	–	–	–
G	*Potentilla erecta*	–	–	o	–	–	–	o	–	–	–	–	–
G	*P. reptans*	–	–	–	–	o	o	o	o	o	o	–	o
G	*Poterium sanguisorba*	o	–	o	o	–	o	o	–	–	–	–	–
G	*Primula veris*	o	–	–	o	–	o	o	o	o	o	–	o
G	*Prunella vulgaris*	–	–	o	–	o	–	o	–	–	–	–	–
G	*Scabiosa columbaria*	–	–	–	+	–	o	o	–	–	–	–	–
G	*Senecio jacobea*	–	–	o	o	+	o	o	o	–	–	–	–
M	*Sonchus arvensis*	–	o	o	–	–	–	–	–	–	–	–	–
M	*S. oleraceus*	–	–	o	o	o	–	o	o	o	o	o	o
M	*Stellaria media*	–	–	–	–	–	–	–	–	–	–	–	o
G	*Succisa pratensis*	–	–	–	–	–	–	o	–	–	–	–	–
G	*Taraxacum laevigatum*	–	–	–	–	–	–	–	–	–	–	–	–
M	*T. officinale*	–	o	o	o	o	o	o	o	o	o	–	o
G	*Thymus drucei*	–	–	–	–	–	–	–	–	–	–	–	–
M	*Tragopogon pratensis*	–	–	–	–	–	–	–	–	–	–	–	–
G	*Trifolium pratense*	–	–	o	+	–	–	o	+	o	o	–	+
G	*T. repens*	–	–	o	o	o	o	o	–	–	–	–	–
M	*Urtica dioica*	–	–	–	–	–	–	o	–	o	o	o	o
G	*Veronica chamaedrys*	o	o	–	o	o	o	o	o	o	o	–	o
G	*Viola hirta*	o	o	–	–	–	–	o	–	–	–	–	–
M	*V. riviniana*	–	–	–	–	–	–	o	–	–	–	–	–
	Total species in quadrat	19	32	42	31	30	31	35	13	10	12	6	17
	Grassland species (G)	13	22	30	14	16	15	17	7	1	0	0	1
	Marginal species (M)	2	7	4	5	8	6	11	2	3	3	0	4
	Woodland species (W)	0	0	0	0	0	0	0	0	2	5	0	5

143

species in secondary woodlands when they were compared to ancient primary woodlands.

When management of grassland ceases many of the woody species are only able to invade during the early years of the resulting succession, so that even-aged stands of scrub are common. Nevertheless a few species are able to invade throughout the sere because they are more successful under conditions of shading and competition. Species with large seeds such as *Quercus* are more likely to be able to establish successfully, while species such as *Prunus spinosa* are able to compete by their habit of suckering. If trees have invaded during the early stages they continue to grow and eventually become dominant, while the shrubs die out or form part of the shrub layer in the woodland. Some species, such as *Juniperus* and *Hippophaë*, are so sensitive to shading that they cannot survive under a closed canopy, while others such as *Daphne laureola* and *Hedera helix* are woodland species and enter only in the later stages. In cases where relatively few trees have invaded with the shrubs, a dense canopy of scrub may form and prevent trees from growing; for example some types of *Crataegus* scrub. These very late stages of succession to woodland are poorly understood, but presumably the bushes will die eventually and gaps will appear in the canopy, thus enabling other species to colonize.

The mosaic situation of patchy scrub, where there are grasslands, 'edge', and woodland species, may persist for variable periods of time especially where the soil differs locally in depth and nutrient status, and where there is interference by man in cutting, burning or by the grazing of domestic animals. At low grazing intensities mosaics of scrub and grassland develop as groups, or individual plants, grow out of reach of the animals, while in other areas the grazing is localized. Where grazing intensity and frequency of cutting and burning has varied, as on some commonlands, 'thicket scrub' may develop (Salisbury, 1918). This has very densely crowded woody stems, shading is often intense and there are few or no field-layer species. When existing scrub is very heavily grazed and trampled the grassland is likely to increase in extent, and if the heavy grazing persists the trees and bushes will eventually die of old age or as the result of damage. This is called retrogressive succession. In those complex situations where there has been variable grazing intensities and other interference, the woody plants present are likely to reflect these events in the age structure and species composition as was shown by Roughton (1972) for a range in Colorado.

Microclimate changes during succession because of the growth of the woody plants, and this brings about corresponding changes in the fauna and flora. The subject has not been much studied, although Stoutjesdijk

(1961) has shown that there was a decrease in the amount of radiation reaching the soil in dune scrub as the leaf canopy thickened, and a reduction in the wind velocity. A generally more equable climate is produced because of these two factors. On still bright nights, temperatures within the closed scrub were as much as 6–7°C higher than in the open. Gradients within the closed scrub were such that the lowest temperatures were near the ground, while the humidity gradients were the reverse depending on type of litter and field layer.

Animals of Scrub Communities

In the scrub ecosystem some animals are particularly important because of their interactions with the plants. These key species influence the establishment of the woody species and the balance between grassland and scrub. Grazing and browsing species are obviously important, for in high numbers they may completely prevent the growth of woody species, and in low numbers mosaics of scrub and grassland may be created. Even where domestic animals are absent, grazing by wild species may prevent the woody species from establishing in some areas. Rabbit grazing is responsible for the characteristic zoning of vegetation around their warrens, which are often established initially under the cover of scrub. Near-by shrubs such as the palatable *Crataegus* may be severely damaged by ring-barking. In Scotland deer are important in influencing the regeneration of woody species. The susceptibility of juniper to damage by red deer depends to a considerable extent on the frequency, amount and persistence of the snow cover in winter (Miller, 1971). Severe damage to the vegetative parts of scrub plants by invertebrates can also occur, and for example *Galerucella viburni* (Col. Chrysomelidae) sometimes causes defoliation of *Viburnum lantana*.

The production and dispersal of the seeds of the woody species has already been considered (p. 135) but the interrelations of the plants and the animals can be further emphasized. It is likely that many species have been evolving in relation to each other, in the same way as the pine squirrels and conifers studied by Smith (1971). He found that characteristics of the plant's reproduction, such as the numbers of seeds per cone and the time of shedding, were selectively influenced by the feeding behaviour of the squirrels. In Europe, the nomadic behaviour of the fieldfare (*Turdus pilaris*) is thought to be adapted to finding the locally abundant sources of berries in winter (Ashmole, 1962). The different species of thrushes are adapted to feeding more on some fruits than others. The song thrushes (*Turdus philomelos*), for example, seem to be specialists on the berries of

Taxus and *Sambucus*, which ripen in the early autumn (Hartley, 1954). Predation on seeds by invertebrates is also important in determining how many seeds survive both before and after dispersal (Janzen, 1971). For example out of 114 seeds examined from an old juniper bush, 44 were found to be killed by *Megastigmus bipunctatus* (Hym. Torymidae), 63 seeds were dead for unknown reasons and only seven were apparently viable. The overall effects of these seed-feeding invertebrates on the scrub ecosystem are not known. The significance of pollinating insects is also uncertain; they do not seem to be limiting in present-day semi-natural vegetation, although it is well-known that they are important in the pollination of fruit trees (Moreton, 1969).

The key species of ecosystems are also likely to be important in schemes for the biological control of weed plants. In addition to their potential to influence the competitive capacity of the weed species, these agents of biological control must also be host specific. Thus, for the control of *Rhamnus catharticus* in Canada, two species of Geometrid moth, which caused defoliation, were suggested for introduction by Malicky, Sobhian and Zwölfer (1970) after they had surveyed the host ranges, feeding sites and phenology of insects associated with European Rhamnaceae. Similarly, for *Ulex* control in New Zealand Schröder and Zwölfer (1969) suggested the introduction of two foliage-feeding species, a moth and a weevil, in addition to the seed-feeding weevil *Apion ulicis* which was already present.

Since scrub vegetation is constantly changing during succession, the habitats available to animals alter also, and species associated with grasslands are slowly lost while others adapted to scrub and woodland appear. This subject has been very poorly studied, apart from the changes during succession from sandy beaches through pioneer communities to a sugar maple/beech forest climax in the Chicago area. This work has been summarized briefly by Allee *et al.* (1949). Some species were found to be tolerant and to occur widely throughout these seral changes, while others were much more limited and were adapted to one particular stage. In the description of the scrub fauna which follows some attempt is made to consider this background of seral change to try to account for the narrow ecological amplitude of some species, but there is little information on most aspects, so that, of necessity, this is an incomplete description.

Food chains based on plant species

Many species, mainly of invertebrates, are found in food chains dependent on plant species, so that as the plant species change during succession this is followed by changes in the food chains as one species of plant is lost and another appears (Richards, 1926). Invertebrates with strict specificity for one species of food-plant will occur therefore only during the seral stage at which that plant flourishes. For example, the sawfly *Monophadnoides puncticeps* feeds on *Poterium sanguisorba* (Benson, 1952), a plant which is lost when shading by bushes becomes intense. Some predatory and parasitic species may be host-specific also, and if their prey species is itself specific to a plant, then these species too are lost if seral changes eliminate the plant. The relationships may be even more complicated. Fisher (1970) showed that some species of Chalcids which attacked the weevils of the genus *Apion* associated with common flowering plants were sometimes more host-plant specific than were their hosts. They attacked host species when they were present in one species of *Rumex* but not in another. Most species of predators are less specific in their requirements, however, and feed on a variety of hosts and so are less likely to be influenced by the loss of one or two plant species.

Taxonomic difficulties with the parasites are one of the main reasons why all the species found in food chains on particular plant species are so incompletely known. In the very detailed investigations of the fauna of *Sarothamnus scoparius*, the Scotch broom, summarized by Waloff (1968) some 70 parasites and 60 common predatory species were found to be associated with the 23 phytophagous species. The fauna could be divided into the seed- and pod-feeding species, the sap-sucking species, the stem miners, and the defoliaters. Many of these species have been studied very carefully. The complex relationships that occur can be illustrated by the fauna associated with the seed pods, where 23 species were found (Fig. 6.5). Parasites and hyperparasites depended on the five phytophagous species and there was also an inquiline associated with the galls caused by *Asphondylia sarothamni* (Dip. Cecidomyiidae). Parasitism could be extremely high at times; for example over 80% of the larvae of the weevil *Apion fuscirostre* might be killed by the parasites. Competition occurred between the inquiline *Trotteria* (Dip. Cecidomyiidae) and the original gall-former *Asphondylia*, and probably also between the weevils *Bruchidius* and *Apion* on the seeds. In addition intraspecific competition was recorded; *Bruchidius* larvae for example competed in one year for a reduced number of broom pods because 62% of the pods had been killed by frost. Between 29 and 37% of the overcrowded larvae that hatched from the eggs laid on the remaining pods died. Similar examples could be given for other members of the broom fauna. The arthropod predators were studied mainly by the use of the precipitin test. It was found that although few of the

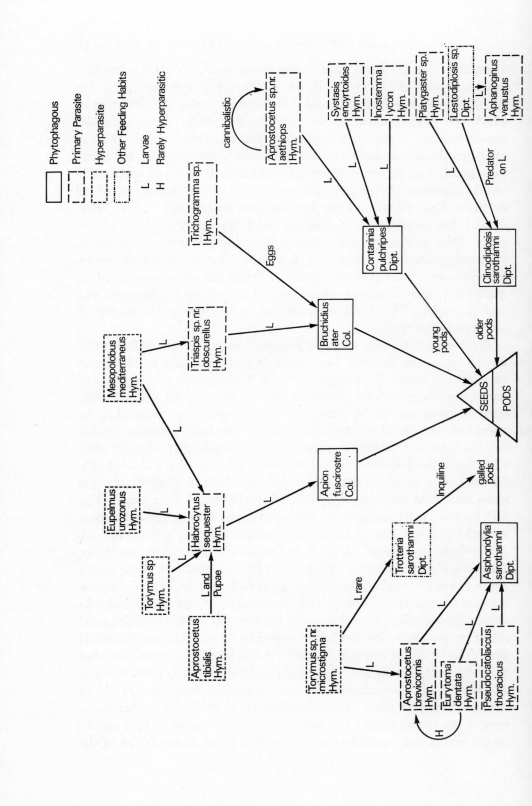

predatory species were selective in their diets, as only a low proportion of any one species was feeding on any particular host, the combined action of all the predatory species could destroy an enormous proportion of the prey population. The relationships of the broom fauna to the bird predators were also investigated. It was found that the predation by the chaffinch (*Fringilla coelebs*) and the blue tit (*Parus caerulus*) was one of the two most important factors checking the population of the broom aphid, *Acyrthosiphon spartii*, at its peak. These birds also fed on the Scolytid beetles within the stem mines in autumn and winter.

Another less detailed study has been carried out on the food chains associated with the insects living on *Viburnum lantana* (Side, 1955). Many interrelationships were noted (Fig. 6.6). The aphid, *Ceruraphis eriophori*, for example, was fed on directly by various predators, its honey dew was used by ants, and the debris left in the leaf curl gall caused by the aphid was used by a number of other insects.

Studies of all the species in food chains associated with other shrub species have not been carried out. However it is possible to compile lists of phytophagous species associated with the shrubs as these species are much better known than the parasitic species, and are also more specific to the plants than most predators. On Table 6.2 estimates of the numbers of insect species feeding on the common shrubs in Britain have been made. A species has been included if it feeds on up to eight plant genera, so that the list is mainly of the monophagous and oligophagous species. Considerable differences in the numbers of species on each genus are found, *Buxus* having only four in comparison with 230 on *Crataegus*. Biochemical and physical barriers are likely to account for some of these differences as species groups evolve in relation to taxonomic groups of plants. For example, Eastop (1972) found that on the 400 species of Coniferae there were 365 species of aphids, of which 270 were Cinarini and 46 Adelgidae, both groups entirely confined to the conifers. The remaining aphid species could be associated with other plant groups where near relatives are found. These species indicate that host transfer has occurred, and in fact Eastop considered that although aphids were 99·9% host-specific, much of the evolution of the group had involved the acquisition of new hosts. The abundance of hosts is often important in the transfer of species as Southwood (1961) found for species associated with trees in Britain. The numbers of species were proportional to the tree's recent abundance, and trees that had only been in Britain for short periods of time had smaller faunas than those of very long duration in this country. Actual acquisition of new hosts has been observed in this century in some Mirids which have

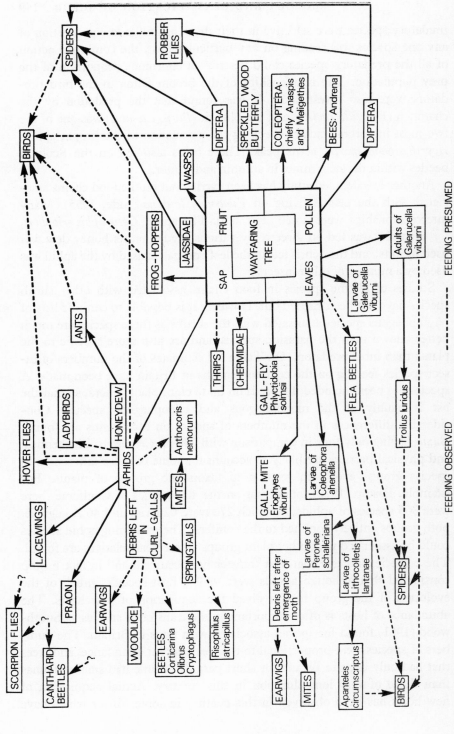

Fig. 6.6 Food chains on *Viburnum lantana* (Wayfaring tree). (*After* Side, 1955.)

—— FEEDING OBSERVED

----- FEEDING PRESUMED

transferred from *Salix* to the abundant *Malus* (apple) in orchards (South-wood, 1972). Both the evolutionary and abundance factors are likely to have played a role, therefore, in determining the high numbers of species recorded as feeding on the Rosaceae (Table 6.2).

Table 6.2 The approximate number of species of the principal phytophagous insect groups associated with genera of scrub woody plants

Plant genus	Macrolepidoptera	Microlepidoptera	Heteroptera	Homoptera – Aphididae	– Psyllidae	– Coccidae	– Aleyrodidae	Hymenoptera – Symphyta	– Cynipidae	Coleoptera	Thysanoptera	Diptera – Cecidomyiidae	– Agromyzidae	Total
Acer	7	12	3	7	0	2	0	3	0	2	1	4	0	41
Buxus	0	0	2	0	1	0	0	0	0	0	0	1	0	4
Clematis	13	0	0	0	0	0	0	0	0	2	0	0	3	18
Corylus	28	28	16	4	0	1	2	10	0	9	3	6	0	107
Crataegus	88	53	14	12	3	5	1	12	0	33	2	6	1	230
Euonymus	2	9	0	2	1	5	0	0	0	0	0	0	0	19
Ilex	3	2	0	2	0	2	0	0	0	3	0	0	1	13
Juniperus	5	8	6	1	0	2	0	1	0	2	1	1	0	27
Ligustrum	17	10	0	1	0	1	0	2	0	1	1	2	0	35
Lonicera	20	12	0	2	0	0	1	5	0	1	1	2	4	48
Malus	34	43	15	11	3	4	0	2	0	16	2	2	1	133
Prunus	57	43	5	6	2	4	0	13	0	17	4	6	0	157
Rhamnus	6	11	2	2	3	1	0	0	0	0	0	2	0	27
Rosa	27	25	2	10	0	4	0	22	10	3	1	3	0	107
Rubus	45	18	3	4	0	2	3	16	1	6	3	5	1	107
Sambucus	4	3	1	4	0	0	0	2	0	1	1	2	1	19
Sorbus	2	4	1	1	0	3	0	13	0	11	0	1	0	36
Taxus	2	2	0	0	0	1	0	0	0	0	0	1	0	6
Thelycrania	6	7	0	2	0	1	0	1	0	0	0	1	0	18
Ulex	9	12	6	2	2	2	0	0	0	12	7	0	0	52
Viburnum	3	3	0	4	0	1	1	3	0	1	0	1	0	17

The probability of finding all the species of insects listed on Table 6.2 on a shrub genus on any one site is very low for a number of reasons. Thus the distribution of an insect species geographically does not necessarily corres-pond to the full range of the plant, but is superimposed on it, often due to different limiting factors of climate. The fauna of juniper, for instance, varies from north to south: stands of the plant on the chalk in England will have *Dichrooscytus valesianus* (Heteropt. Miridae), *Cyphostethus tristriatus*

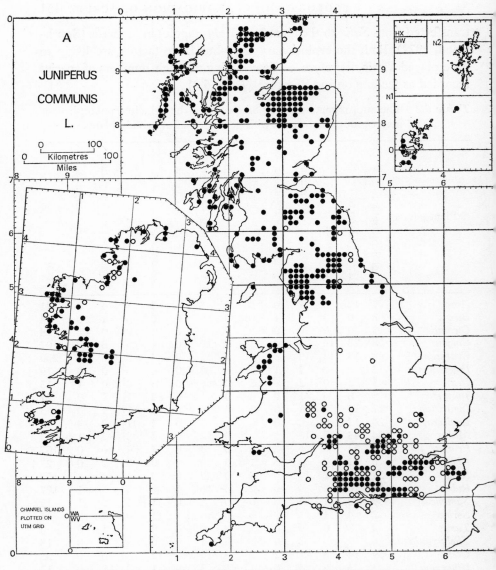

Fig. 6.7 A–C The distribution of two species of *Thera* (Lepidoptera : Geometridae) and of their food-plant, *Juniperus communis* (juniper) in Britain on a 10 kilometre grid.

(A) The distribution of *Juniperus communis* in southern England according to Ward (1973); northern England, Wales, Scotland and Ireland according to Perring and Walters (1962). For the distribution of the sub-species see Perring (1968).

● Post-1965 in southern England; post-1930 elsewhere.
○ Pre-1965 in southern England; pre-1930 elsewhere.

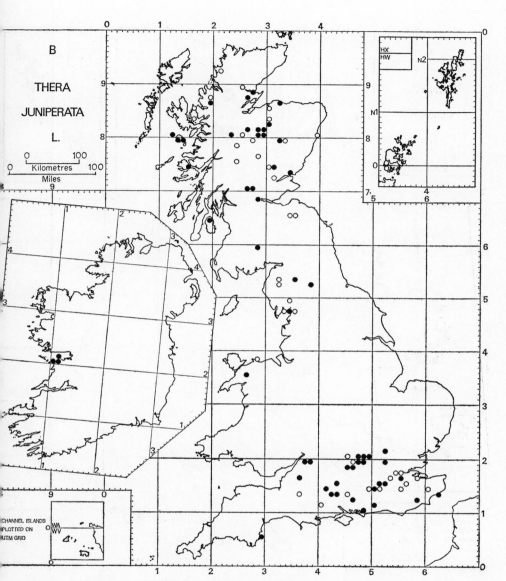

(B) The distribution of *Thera juniperata* (juniper carpet moth.) *T. juniperata juniperata* occurs as far north as the Lake District, *T. juniperata scotica* occurs in Scotland, while *T. juniperata orcadensis* is found in the Orkneys. (A very few casuals have been recorded at light in the Midlands, and may possibly be associated with ornamental junipers in gardens.)
● Post-1960   ○ Pre-1960.

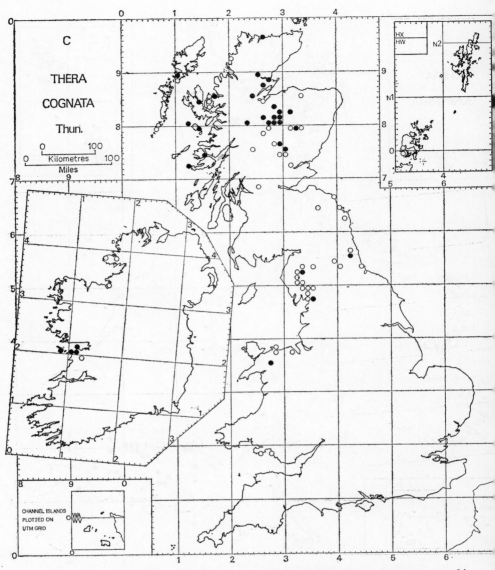

(C) The distribution of *Thera cognata* (chestnut-coloured carpet moth).
● Post-1960   ○ Pre-1960.

154

(Heteropt. Acanthosomidae), *Schmidtiella gemmarum* (Dip. Cecidomyiidae) and the southern typical form of the juniper carpet moth, *Thera juniperata juniperata*. In Scotland the two Heteroptera will be absent, although on Speyside there may be another Mirid *Zygimus nigriceps* (Waterston, 1964). The Cecidomyiid will be replaced by another species, *Oligotrophus juniperinus*. *Thera juniperata* sub-species *scotica* will be present, together with the chestnut-coloured carpet moth, *Thera cognata*. The distribution of the two species of Lepidoptera and of the main host plant, *Juniperus communis*, is shown on Fig. 6.7, A,B,C. Parasites and predators too will differ geographically, and it has been suggested that the distribution patterning of the larch sawfly is due to the changing population of parasites found in different areas of its holarctic range (Turnock, 1972). In areas where the sawfly is never a pest there is a diverse environment and rich parasite complex, while where a permanent type of severe damage is caused, there is an impoverished parasite complex, no effective specific parasites and some interactions of the populations with food availability. Situations where periodic outbreaks occur can be associated with an intermediate complex of parasites and one very effective specific larval parasite.

The different habitats of the host plants may also influence the associated insects within their broad geographical range, some species being found wherever the host occurs, while others are much more restricted. Welch (1972) found, for example, that the larvae of *Hermaeophaga mercurialis* (Col. Chrysomelidae) were limited by the need for a light, preferably calcareous, dry soil, and so occurred on *Mercurialis perennis* on the chalk but not in woods on clay. The black hairstreak butterfly, *Strymonidia pruni*, is restricted in its preference for old *Prunus spinosa* in ancient woodlands in the Midlands. This may well be related to its poor powers of dispersal as well as to some preference for microclimate. Taking a broad view of the fauna associated with *Crataegus* throughout the world, Wellhouse (1922) thought that the plant could be considered as occurring in five main community types because of the combinations of the associated insects with other hosts and habitats. The five types were open woods, deforested areas, grazing lands, stream banks and fence rows.

The genetic background of adaptation to environment and the host plant and insect interactions is immensely complex. Genetic differentiation between geographical populations is well known and has been reviewed by Gould and Johnson (1972). Oliver (1972) studied four species of Lepidoptera and found that there was some level of genetic incompatibility between all the populations compared, and that where there was a fairly linear series geographically there was a gradual accumulation of genetic incompatibility and phenotypic differentiation with distance. The sub-species of the juniper-feeding moth *Thera juniperata* are found in different geographical areas of Britain (Fig. 6.7B) and these are presumably adapted to the local conditions. Little is known about this, except that the sub-species *scotica* pupates and emerges earlier than the typical form, as well as having a slightly different wing pattern. To illustrate the complexity of the genetic interaction between the host plant and the insect, studies by plant-breeders on resistance can be cited. Keep (1972) illustrated some aspects of this problem in studies on the genetics of the wild raspberry, *Rubus idaeus*. This species shows polymorphism for a number of major genes related to resistance to the powdery mildew and also to the *Rubus* aphid, *Amphorophora rubi*. It was complicated by the fact that there were a number of races of the aphid which behaved differently in relation to the plant.

Animals and the life cycles of the plants

The response of an insect species to a host plant may vary according to the age of the plant, and during succession shrubs often pass through their life cycle of vigorous young growth, maturity and senescence. A decline in population growth and an increase in resistance to invasion of insects is often associated with older plants.

This is thought to be related to the nutrient status of the plants, for example, the presence of amino acids in the case of some aphids (Van Emden and Way, 1972). In some species, such as the fruit tree red spider-mite, *Panonychus ulmi*, diapause may be induced and there is also evidence that growth and reproduction are influenced in some insects and mites. When a number of species are associated with a host plant most prefer young growth but a few species may be more successful on the old plants as Waloff (1968) has shown for *Sarothamnus*. This shrub lives for only some 10–15 years, but most of the insects arrive within the first two. During the early stages of growth the young shoots are more favourable to Psyllids, which have denser populations, but as the plant ages these populations fall. The fecundity of several of the broom species is reduced on the older bushes, as for example *Phytodecta olivacea* (Col. Chrysomelidae). On the other hand, the Scolytid beetle *Hylastinus obscurus* was more numerous on the old bushes which were sometimes killed by its attack. Dead branches were attacked by another Scolytid *Phloeopthorus rhododactylus*. In general, Waloff thought that succession on the bushes was influenced more by the changing densities of populations than by the complete absence of species.

When flowering and fruiting takes place the insects dependent on these microhabitats are able to invade the sere. Many shrubs produce abundant blossom which is visited by insects such as the flower beetles, *Meligethes*, and other species (Plate 22). Few insects however, feed only on the flowers of one species of shrub throughout their life cycles, apart from some of the flower-dwelling thrips. Fruits and seeds on the other hand do have a specialized and specific fauna in many cases. *Juniperus*, for example, matures and begins to fruit at about 10–15 years of age in southern England, and only at this stage do the juniper shield bug, *Cyphostethus tristriatus*, and the seed chalcid, *Megastigmus bipunctatus*, both of which feed exclusively on juniper berries, begin to colonize the sere. Similarly there is a group of species associated with rose hips (Eichhorn, 1967) although in this case colonization is earlier because the fruits appear within three to four years.

The formation of wood by trees and shrubs allows animals adapted to this habitat to establish themselves. Hollow woody stems, for example, may be used by Hymenoptera Aculeata for nesting. Some species are found almost exclusively in the dead stems of *Rubus*, some preferring stems located in sunny situations while others prefer the shade (Danks, 1971). Scolytid beetles attack both living and dead wood. These tend to infest older trees or those with branches dying through disease, although some of the species attacking fruit trees have also been found on healthy trees (Massee, 1954). In the succession of insects found on the different stages of dead and dying wood, Scolytid bark beetles and their associated parasites are often amongst the first species to invade. Many are thought to spread the ambrosia fungi which accelerate the rotting process. However, many other species of fungi soon invade dead wood and once the bark has been loosened a rich community of arthropods develops, some of which feed on the fungi as well as on the decomposing wood. Millipedes, centipedes, spiders, pseudoscorpions, earwigs, collembola and various beetles are common (Elton, 1966; Stubbs, 1972).

Lichens and algae, which grow on bark, also colonize wood in scrub, bringing with them their own special fauna. Psocoptera are often very numerous on the older shrubs, many of them living on the bark flora, although others are associated with the epiphytic plant life on the foliage. Psocoptera species are not strictly host-specific although some are found more commonly on some trees than others; for example, there is a marked difference between conifers such as *Juniperus* and deciduous species such as *Crataegus*. The Psocoptera also have their associated hymenopterous parasites and other enemies (New, 1971). Other insects characteristic of lichen-covered trees and bushes are the bugs of the family Microphysidae.

Many other species feed on the bark flora, particularly the Collembola and Oribatid mites, as well as a number of camouflaged Lepidoptera larvae such as *Laspeyria flexula* and the case bearer *Luffia ferchaultella*.

Changes in plant structure and the fauna

The changing structure of trees and bushes during growth alters the microclimate (p. 144) and provides different spatial arrangements and heights of vegetation. This has a considerable influence on the fauna, providing, for example, shelter in winter for many species, particularly on evergreens such as *Taxus* and *Juniperus*. Psyllidae, some Typhlocybid Auchenorhyncha and Chalcids are particularly common in *Juniperus* in winter.

Some phytophagous insects which are less specific to plant species have been found to be dependent on the height of the plants in grasslands (Chapter 5). Thus as the field layers of scrub stands change in height during succession, the numbers and species of associated insects alter. A number of predatory invertebrates, for example, the spiders, are also affected by the structure of habitats (Duffey, 1962b, 1966). The microclimate as well as stem rigidity is important for web-builders. Thus, *Ulex* is a particularly favourable plant for spiders (Bristowe, 1939).

Air movements are set up near windbreaks consisting of trees and shrubs and flying species are often deposited in the wind shadow (Lewis and Dibley, 1970). Although the subject has been little studied it is obvious that patchy scrub is likely to have similar effects on flying insects. They may concentrate in certain areas, and would be able to fly within a scrub area when it is too windy in more exposed situations.

Birds are also affected by structure, using the shelter of bushes for nest-building as well as utilizing the insects and berries as a source of food. Bevan (1964) described the feeding sites of birds in grassland with thick scrub, and found that five main feeding areas could be designated, the trees, shrubs, herbs, the ground, and the air. The song thrush (*Turdus philomelos*), robin (*Erithacus rubecula*), dunnock (*Prunella modularis*), and chaffinch (*Fringilla coelebs*) preferred the ground while the blackbird (*Turdus merula*), redwing (*Turdus iliacus*), fieldfare (*Turdus pilaris*),various tits, greenfinch (*Carduelis chloris*), redpoll (*Acanthis flammea*), bullfinch (*Pyrrhula pyrrhula*), whitethroat (*Sylvia communis*) and willow warbler (*Phylloscopus trochilus*) feed mainly in the canopy of shrubs and trees. Breeding birds may be present in scrub in high numbers. Williamson (1967) for example found 36 species with about 540 pairs on 83 ha of open scrub of *Crataegus* on the Ivinghoe Hills in Bucks (see also Chapter 5).

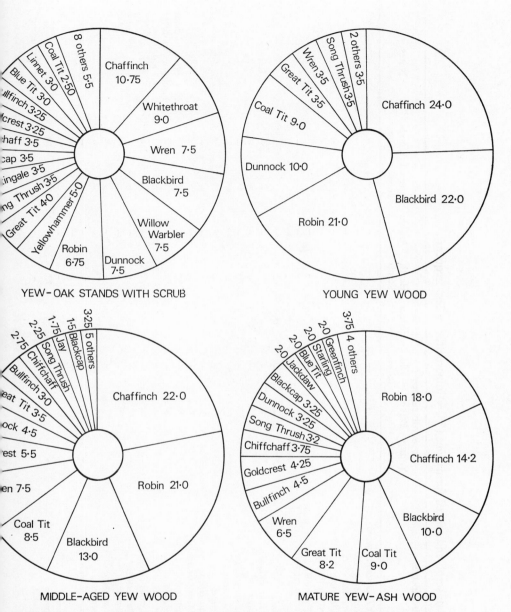

Fig. 6.8 Numbers of each breeding bird species (as a percentage of the total to the nearest 0·25 %) in different stages of the yew-wood succession at Kingley Vale National Nature Reserve, Sussex during 1967–71. (*After* Williamson and Williamson, 1973.)

Table 6.3 Birds of scrub and related formations (** = important, * = less frequent),

Bird species	Scrub canopy			Hedges	Heath scrub	Arable	Scattered trees	Wood-land	Notes
	Young open	Patchy	Old closed						
Grasshopper warbler *Locustella naevia*	**	—	—	—	—	—	—	—	
Stonechat *Saxicola torquata*	**	—	—	—	**	—	—	—	
Whinchat *Saxicola rubetra*	**	—	—	—	—	—	—	—	
Willow warbler *Phylloscopus trochilus*	**	—	—	—	—	—	—	—	
Meadow pipit *Anthus pratensis*	**	*	—	**	—	—	**	*	
Dunnock *Prunella modularis*	**	*	—	**	—	—	**	*	
Yellowhammer *Emberiza citrinella*	**	—	—	**	—	—	**	—	
Partridge *Perdix perdix*	**	—	—	**	—	**	—	—	
Quail *Coturnix coturnix*	**	—	—	—	—	**	—	—	
Dartford warbler *Sylvia undata*	**	**	—	—	**	—	—	—	Rare
Cuckoo *Cuculus canorus*	**	**	—	**	**	—	—	—	
Nightjar *Caprimulgus europaeus*	**	**	*	—	**	—	**	—	
Pheasant *Phasianus colchicus*	**	**	**	—	—	**	**	**	
Kestrel *Falco tinnunculus*	*	*	—	—	—	—	**	*	
Cirl bunting *Emberiza cirlus*	*	*	—	**	—	—	—	—	
Green woodpecker *Picus viridis*	*	*	—	—	—	—	**	*	
Marsh tit *Parus palustris*	*	*	—	—	**	—	**	**	⎫ Nest often in dead
Willow tit *Parus montanus*	*	*	—	—	**	—	**	**	⎬ birch, occasionally elder
Red-backed shrike *Lanius collurio*	—	**	—	—	**	—	—	—	Rare
Turtle dove *Streptopelia turtur*	—	**	—	*	—	—	—	*	
Blackbird *Turdus merula*	—	**	—	**	—	—	*	**	
Whitethroat *Sylvia communis*	—	**	—	**	**	—	—	*	
Linnet *Acanthis cannabina*	—	**	—	**	**	—	—	—	
Reed bunting *Emberiza shoeniclus*	—	**	—	**	—	**	—	—	Damp areas
Magpie *Pica pica*	—	**	—	**	—	**	—	—	
Robin *Erithacus rubecula*	—	**	**	*	**	—	—	**	
Redpoll *Acanthis flammea*	—	*	—	*	**	**	**	**	
Goldcrest *Regulus regulus*	—	*	*	—	—	—	—	**	Conifers including juniper

160

Species						
Goldfinch *Carduelis carduelis*	–	*	–	–	–	**
Redstart *Phoenicurus phoenicurus*	**	*	–	–	–	**
Woodpigeon *Columba palumbus*	**	**	**	–	–	**
Hawfinch *Coccothraustes coccothraustes*	–	*	–	–	–	**
Greenfinch *Carduelis chloris*	–	*	–	**	–	*
Chaffinch *Fringilla coelebs*	–	**	–	–	–	**
Jay *Garrulus glandarius*	–	**	–	–	–	*
Longtailed tit *Aegithalos caudatus*	–	**	–	*	–	**
Wren *Troglodytes troglodytes*	–	**	–	**	–	**
Song thrush *Turdus philomelos*	–	**	–	**	**	**
Nightingale *Luscinia megarhynchos*	–	**	–	–	–	**
Blackcap *Sylvia atricapilla*	–	**	–	–	–	**
Garden warbler *Sylvia borin*	–	**	–	–	–	**
Lesser whitethroat *Sylvia curruca*	–	**	–	*	–	**

161

Williamson and Williamson (1973) studied the bird communities of the yew wood succession at Kingley Vale, and found the richest diversity of birds in the groups of older yew and oak trees with fringing scrub. The dense young yew wood was poorer in species and numbers than the middle-aged wood, while the mature yew–ash wood had more species, perhaps because of the greater availability of hole nest sites (Fig. 6.8). In oakwood successions birds were found to be most numerous in the mature stands with associated scrub (Hope-Jones, 1972). Stuttard and Williamson (1971) found that the numbers of nightingales (*Luscinia megarhynchos*) varied according to age of coppice. Coppice of between 5–8 years was most suitable and the density of oak stands also influenced the distribution of breeding territories.

Table 6.3 has been compiled to show the bird species commonly associated with scrub and related formations which have been divided into early, middle and late stages of succession. Many species found commonly in hedges and in arable areas are also shown on the Table while others are found mainly in *Ulex* scrub on heathlands. The red-backed shrike (*Lanius collurio*) once common in scrub, is now found locally on heathlands in the south (Ash, 1970). Areas with scattered trees are normally required for nesting or song posts by some species (also indicated on the Table) although some, such as the sparrow hawk (*Accipiter nisus*), will nest in scrub if there are no trees available.

7 Some concepts of grassland management

Introduction

Semi-natural grasslands in temperate regions consist of an intimate mixture of grasses, broad-leaved plants and mosses. One of the important ways in which the biological spectrum of plants and animals in grasslands may be controlled or changed is by management, and the purpose of this and the following two chapters is to discuss what should be done in order to develop swards of different types. However, first it is necessary to describe some characteristics of grassland plants, in order that we may formulate general rules which are applicable to the conservation management of most temperate grasslands.

The variation in the growth-form of grassland plants can best be demonstrated by classifying them according to a system developed by Raunkiaer (1934). He described seven main groups which he called *Life-forms*, distinguished according to the position of the overwintering buds (Table 7.1).

Hemicryptophytes, which have the overwintering bud and growing

Table 7.1 Percentages of different life-forms found in samples of chalk grassland, dry acidic grassland and meadow grassland (Analysis based on 182, 53 and 300 species respectively)

Life-form	Position of overwintering bud	Chalk grassland %	Acid grassland %	Neutral grassland %
Chamaephytes	Soil surface to 25 cm	7·6	20·8	7·2
Hemicryptophytes	At soil surface	67·6	54·6	70·3
Geophytes	Below soil surface	16·5	5·6	13·5
Therophytes	As seed	8·2	18·8	8·6

point at, or near, the ground surface, account for about two-thirds of plants in temperate grasslands. This group includes not only perennial grasses and sedges, which in many types of grassland make up the greater proportion of species, but also numerous dicotyledonous herbs. Some of these, such as *Asperula cynanchica* and *Galium verum*, have a prostrate habit of growth with small basal leaves. Others have a semi-rosette habit (*Anthyllis vulneraria*) or a rosette habit (*Hieracium pilosella*). All are adapted by habit of growth and position of their apical meristem to withstand grazing and, hence, cutting or mowing.

Another type of plant found in grasslands is the geophyte, which passes the winter as an underground tuber or modified stem. Chalk grassland contains 16·5% geophytes, most of which are orchids (Plate 23). Neutral grasslands have 13·5% of geophytes but in acidic grasslands which are deficient in orchids the proportion is only 5·6%. Chamaephytes form less than 20% of the life-forms of moist, temperate grasslands but this element becomes of greater importance in the warmer, drier parts of the world. In these regions dwarf shrubs, often protected from grazing by spines or by the presence of unpalatable aromatic compounds, are important constituents of so-called grasslands. Annuals are not important components of closed communities, a characteristic feature of mesic grasslands, but they make a significant contribution to the more open grasslands found on nutrient-poor soils.

Adaptation to Frequent Defoliation

Plants which are frequently or intermittently defoliated must be able to regenerate new tissue if they are to survive in grassland. Grasses produce fresh green material by tillering, the form of which is characteristic for each species, although this may be modified by other factors in the environment. Langer (1956) has suggested that the grass plant is best considered as a dynamic aggregate of short-lived plantlets or tillers, the perennial habit being secured by the overwintering of vegetatively produced shoots which themselves rarely survive the subsequent season. Moreover, plants which are frequently defoliated must be able to synthesize and translocate reserve food substances to storage regions, so that after defoliation or dormancy the plant may produce fresh vegetation capable of photosynthesis. Many studies (May, 1960; Aldous, 1930) have shown that the major reserves in grasses are soluble carbohydrates (sucrose, glucose, fructose, fructosan) and starch. The amount of these reserves stored in the roots fluctuates throughout the year, generally being high in summer and low in spring. As might be expected, the time and intensity of defoliation affects root development, level of food reserves and ultimately plant vigour and competitive ability.

The response of broad-leaved plants to defoliation has received scant attention from agriculturists or physiologists. Nevertheless, the same sort of principles seem to apply as were discussed for grasses, although the morphology and growth of dicotyledons are considerably different. Thus herbs in general do not tiller, but they do reproduce vegetatively by the growth of lateral buds. In many rosette plants, the young leaves are situated in the centre of the plant stem at ground level, protected from grazing by their position and possibly by the presence of hairs. Many of the perennial herbs of grassland have long tap roots and stout underground stems in which food reserves are mobilized and used for leaf production during the period of the year when conditions for maximum production are most favourable.

Perennial Habit and Longevity

An important, but often overlooked, characteristic of grasslands in temperate regions is that more than 90% of the species of plant are perennials, many having a long life-span. The significance of the perennial habit has not always been understood because there have been few studies in western Europe of the dynamics of permanent grasslands, for example longevity or frequency of seed production.

Most grasslands in temperate regions are closed communities with a stratified arrangement of leaves and flowers leaving little or no bare ground. Flower and seed production by perennials varies considerably, the number of propagules and frequency of production being a characteristic of each species. However, these characteristics are also influenced by climatic conditions. In some cases, there may be a direct relationship between a particular climatic feature and viable fruit production, as Pigott (1970) has suggested for *Cirsium acaulon*. On the other hand, the relationship between plant performance and climate may be complex; the studies of Tamm (1948, 1972) and Wells (1967) on species of orchids illustrate the difficulty of identifying the many factors which combine to induce a plant to flower, or to remain vegetative. Furthermore, the production of propagules can be greatly affected by management. For example, cutting or grazing at a time when a species is in bud may effectively prevent it from producing seed in that year. This applies particularly to monocarpic species but less so to those perennials which regularly reproduce by vegetative means.

Studies on meadow plants in Russia (Rabotnov, 1950, 1960, 1969) have

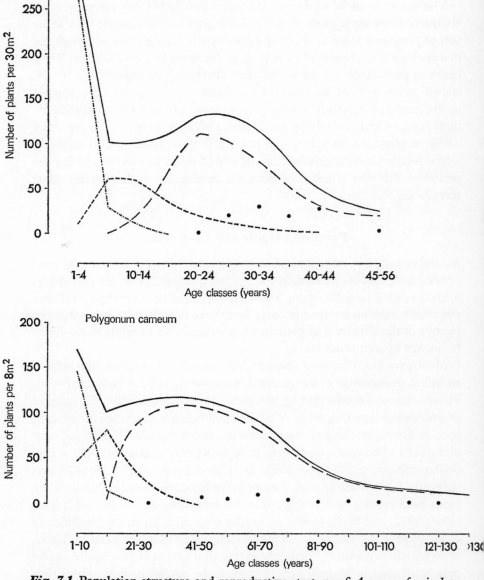

Fig. 7.1 Population structure and reproductive strategy of *Anemone fasciculata* and *Polygonum carneum*. (———) total population; (———) mature vegetative plants; (- - - -) semi-mature plants; (—·—·—·) juvenile plants; (●) reproductive plants. (*Redrawn from* Harper and White, 1971, *originally based on data of* Rabotnov, 1945.)

demonstrated that in closed communities such as meadows, the longevity of some of the constituent species may exceed 130 years, some not flowering until they are at least 20 years old. His work has shown that in any closed community there is likely to be a distribution of plants in different age classes. There will be seedlings, juveniles, immature adult plants, reproductive plants, vegetative adult plants and senescent plants of great age. Data on *Anemone fasciculata* and *Polygonum carneum* which illustrate this point are shown in Fig. 7.1.

The establishment of perennials in grassland by seed undoubtedly occurs but little is known of the environmental conditions which are necessary. Studies of *Anemone pulsatilla* in calcareous grasslands, by Wells and Barling (1971), have shown that establishment of this species from seed is an extremely rare occurrence. The existing colonies of this plant would appear to be relics of older populations which date back to earlier times when the climate was moister and the system of land management provided suitable habitat conditions for seed germination. Vegetative reproduction has enabled this species to maintain itself at old sites, but there is no evidence of it having spread to new ones over the past two hundred years.

Seedlings of other species, such as *Poterium sanguisorba*, are frequently found on small patches of bare ground between tufts of grass in grazed downland. Annuals, such as *Linum catharticum* and *Euphrasia nemorosa*, are also most frequent in short grassland in which cover is not 100%. There is increasing evidence that disturbance of the vegetation cover and the creation of small areas of bare ground are important for seedling development in permanent grasslands. The role of rabbits, moles and earthworms in the grassland ecosystem may be of special significance in this connection.

Phenology

It is now widely accepted that the management of complex communities must be based ultimately on a knowledge of the biology of each component species. This ideal is unlikely to be achieved within the foreseeable future, except for communities containing few species. However, certain facets of the biology of plants are easier to record than others, for example, time of leaf production, time of flowering and setting of fruit. Fortunately from the management point of view, these are some of the more important attributes of the plant which affect animal/plant relationships in grassland. For this reason it is essential to know something of the phenology of both plants and insects if the management of grassland is to have a scientific basis.

Recent work at the Grassland Research Institute at Hurley has shown that for a wide range of pasture grasses growth is bimodal with a major peak of production from mid-May to mid-June and a lesser secondary peak in August. Growth during the winter months is negligible. Fig. 7.2 shows a generalized growth curve for grasses derived from the Hurley data.

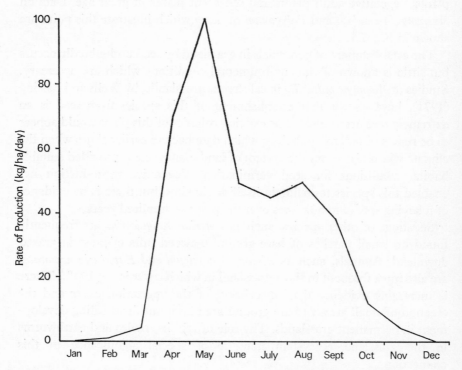

Fig. 7.2 Generalized growth curve for agricultural grasses growing under conditions of adequate nutrients and moisture. (*Based on data given in the Annual Reports of the Grassland Research Institute, Hurley.*)

Morris and Thomas (1972) studied the seasonal pattern of dry-matter production of *Lolium perenne, Cynosurus cristatus, Anthoxanthum odoratum, Festuca ovina* and *Agrostis stolonifera* at five sites situated at different altitudes in the Pennines. They showed that the shape of the dry-matter production curves at the lowest site at Bailrigg (49 m above sea-level) followed the normally accepted pattern of a rapid increase in spring up to a peak, followed by a decline to a lower rate during the summer months (Fig. 7.3). At higher altitudes, for example at Fair Oak Two (302 m above sea level) there was a consistent and highly significant decline in the spring

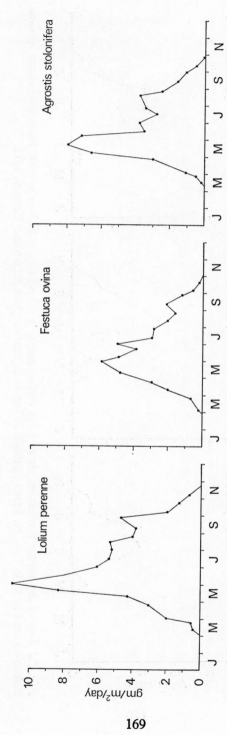

Fig. 7.3 Rate of dry matter production above 3 cm cutting height (g/m² per day) of *Lolium perenne*, *Festuca ovina* and *Agrostis stolonifera* for Bailriggs, N.W. Lancs., altitude 49 m a.s.l. (*Redrawn from Morris and Thomas, 1972.*)

169

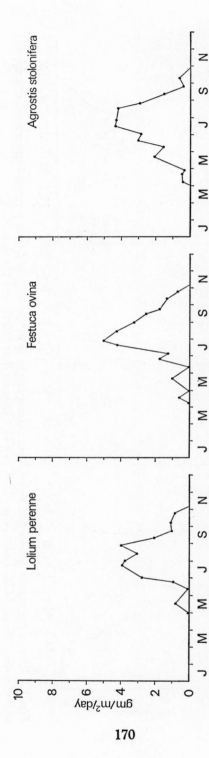

Fig. 7.4 Rate of dry matter production above 3 cm cutting height (g/m² per day) of *Lolium perenne, Festuca ovina* and *Agrostis stolonifera* for Fair Oak Two, N. Lancs., altitude 303 m a.s.l. (*Redrawn from* Morris and Thomas, 1972.)

peak of dry-matter production compared with the lower sites (Fig. 7.4). The greatest depression was recorded for *Lolium perenne*, a reduction from 10·9 g/m² per day at Bailrigg to 0·9 g/m² at Fair Oak Two. Other species were less affected, although their peak production at Bailrigg was also lower than that for *Lolium*. These results show that there are wide variations in both seasonal and total dry-matter production between different environments, and that the performance of different species may not always be in the same relative order.

Dicotyledonous pasture plants have received little attention from agriculturists, chiefly because they are considered to be weeds in grassland and therefore undesirable. However, many are rare or of special interest to ecologists so that they may have an important influence on the selection of grassland nature reserves. There is little information available on the growth cycles of herbs, although phenological observations on a number of chalk grassland species (Fig. 7.5) illustrate the sort of data that are likely to be of use for predicting the effect of management treatments.

Management in Relation to the Life Form, Perennial Habit, Longevity and Phenology of Grassland Species

The pioneer studies of Fenton (1937), Stapledon (1937), Jones (1933) and Milton (1940, 1947) on the management and agricultural improvement of hill pastures and permanent grassland clearly show that the floristic composition of grassland can be manipulated and changed by manuring, grazing and cutting. Some of their findings, for example that the application of basic slag to hill grasslands will increase the clover content, are so well-known that they are now generally accepted. Their work was a significant advance in agricultural research because it showed that the composition of a grassland could be controlled, to some extent, by the grazier. This led to further studies and the publication of many books on grassland management of which those by Stapledon (1939), Davies (1952) and Voisin (1960, 1961) are best known.

Unfortunately, from the point of view of the conservation ecologist, most of their research was done on leys or hill pastures with more emphasis placed on yield and nutritive values than on the dynamics of grassland. It was left to Watt (1938), who studied the rather exceptional grasslands of Breckland, to establish the ecological basis of the dynamics of grassland which culminated in his classical paper on pattern and process in plant communities (Watt, 1947).

Recent studies by Hunter (1962), Nicholson (1971), Nicholson, Paterson

Phenology

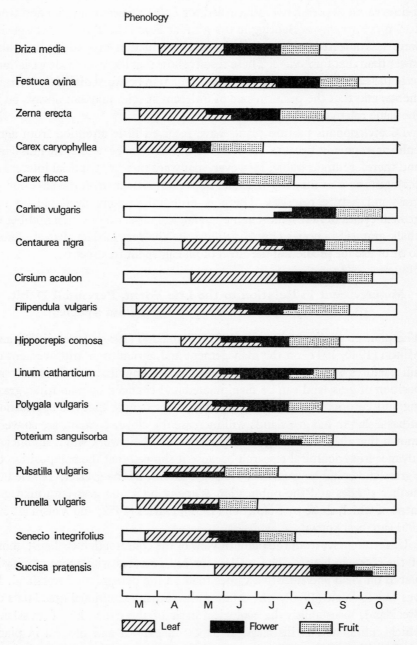

Fig. 7.5 Phenology of some chalk grassland species. (*Data from* Wells, 1971.)

and Currie (1970), Rawes and Welch (1969) and others on hill pastures, and on lowland permanent grasslands by Norman (1957), and Kydd (1964) and Wells (1971) have extended ecological knowledge on the dynamic nature of grassland under a range of grazing and cutting regimes. They have shown that grazed grassland may be remarkably stable so that small-scale patterns of change, due to vegetative growth and spread, do not affect the overall floristic composition of the grassland. Indeed, grazed chalk grassland may be seen as a complex of many species which may change in space and time on a small scale but which nevertheless remains fairly constant in general character provided the grazing regime is maintained. It is not known if this concept can be applied equally well to other grasslands, although there are theoretical grounds for thinking this may be so.

A directional change may be induced in grasslands by particular treatments, leading to an increase in some constituent species and a decrease in others, although the change is rarely continuous and more often a new equilibrium is attained. Grazing and cutting are treatments which may cause such changes but rarely result in the elimination of a species. In contrast, the application of fertilizers, particularly high rates of inorganic nitrogen, cause rapid changes in floristic composition leading to a sward dominated by grasses and with a few other species. Broad-leaved herbs may be completely eliminated after two or three years of heavy nitrogen applications. Of particular importance to the conservationist is the inability of many herbs to re-establish in the thick grass sward which is produced, so that the process is irreversible in contrast to grazing or cutting treatments.

So far we have considered only the botanical aspects of grassland management. It is as well to remind ourselves that the zoological interest of grasslands is high. Grasslands of different floristic composition varying in structure support a great variety of insects, birds and small mammals (Chapters 3 and 5). When making a case to conserve a particular grassland the diversity and richness of plant and animal life are factors of considerable importance. Recent studies on chalk grassland suggest that while floristic richness is most easily achieved by maintaining a fairly short turf managed in order to control the competitive ability of the more productive grass species (Wells, 1971), maximum faunistic richness and diversity is usually associated with much taller vegetation (Morris, 1971b). These two results of recent research do not constitute a contradiction in management recommendations providing one ensures that the objectives of management are clearly stated before management begins.

The objectives, or requirements of conservation, will almost certainly differ from site to site as well as varying in complexity. A distinction must sometimes be drawn between primary and secondary objectives of management depending on ownership and land use. Thus on a National Nature Reserve the primary objective is usually to enhance and maintain the stated scientific interest of the site. On the other hand, the primary objective on a country park, for example, will be to provide facilities for public enjoyment while the maintenance of the ecological interest is secondary.

On sites containing earthworks (Plate 24) or ancient field systems, the primary objective will be to safeguard the site for archaeological study. This may necessitate the removal of scrub, the roots of which may break up soil profiles of importance to the archaeologist. In only a few instances will the distinction between sites be as simple as that outlined above. Where possible the objectives should be defined in order of importance and one should note that the more numerous they are the greater the difficulty may be in reconciling their functions and purpose.

One type of management objective might be to maintain an area of grassland for general richness of plant and animal life without paying special attention to any one species. This form of management 'for general biological richness' is a sound starting point in those cases where detailed knowledge is not yet available as it allows a more sophisticated management programme to be developed at a later stage. Another objective might be to maintain a range of different structures for maximum insect diversity at a site where a grass of high competitive ability grows in some quantity.

These kinds of management, where the primary object is to maintain biological interests, contrast with the problems of local authorities who are responsible for attractive grassland landscapes used for amenity and recreation. Nevertheless the functions of nature reserves and public open spaces are not mutually exclusive and in some cases a good deal can be done to cater for both interests by suitable management treatments (see comments on 'multiple use' in Introduction). The methods available will be discussed in the next three chapters.

8 Management and grazing

The complex relationship between grazing animals and plant communities has long been recognized by pasture workers. Arnold (1962) summarized the more important features diagrammatically (Fig. 8.1) and later (1964) pointed out that in most experimental work making use of herbivores, it is virtually impossible to control more than a few variables at any one time.

Although there is much published information on the agricultural aspects of herbivore-pasture interactions, suitable techniques have yet to be developed for many outstanding problems concerning grazing and wildlife in semi-natural permanent grasslands.

Grazing has three main effects on vegetation: the sward is defoliated, nutrients in the form of dung and urine are returned or removed from the grassland ecosystem and the plant life suffers physical damage by trampling.

Defoliation

Perhaps the most important aspects of grazing which affect the structure and botanical composition of grassland are selection of plants and avoidance of others together with intensity and frequency of grazing.

It has long been known that grazing animals eat certain plants while rejecting others. The famous botanist, Linnaeus, in 1748 (quoted in Tribe, 1950) commented 'This summer I continued my investigations as to which plants are consumed by cattle, which are ignored and which are avoided; this work in my opinion is of fundamental importance both for private owners of livestock and for animal husbandry as a whole'. To investigate this point, Linnaeus offered a total of 618 plant species to sheep, cattle, goats, horses and pigs. He found that in the case of sheep 449 species

175

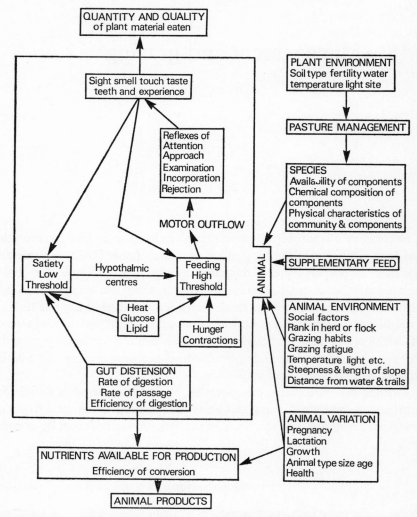

Fig. 8.1 The complexity of soil–plant–animal relationships. (*Source* – Arnold, 1964.)

were always eaten, 32 sometimes, 137 never. Goats and sheep showed least discrimination in selection, a result which is surprising in the light of modern research.

Similar work is reported by Davies (1925) who offered 15 grasses and legumes growing in small plots to sheep and found that clovers were always palatable irrespective of season or condition, while other species, such as *Festuca arundinacea*, were least preferred. Ivins (1952) grew 14 species of grasses and broad-leaved herbs in plots and noted their relative palatability to dairy shorthorn cattle. Clovers were more palatable than other species tested, irrespective of time of year and condition of plants. These observations and experiments show that selection does occur and that many

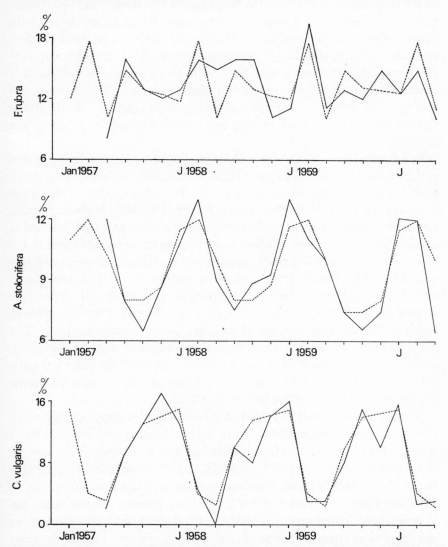

Fig. 8.2 Annual grazing pattern of Blackface ewes at Lephinmore Hill Farm, Argyll, 1957–59. Intake of *Festuca rubra*, *Agrostis stolonifera* and *Calluna vulgaris* measured using faecal analysis. Long-term averages, based on the whole period (1957–59) are shown as dotted lines, with which the results for each particular year, shown as full lines, can be compared. (*Redrawn from* Martin, 1964.)

factors, such as hairiness, amount of fibre and stage of growth, influence the choice of plants.

A more direct and accurate method of studying the diet of herbivores under free-range conditions is the technique of faecal analysis. The epidermal walls of nearly all grasses and some other herbs are thickened with cutin, a fatty substance which is resistant to digestive enzymes so that, providing suitable sampling techniques are used, the fragments of epidermis recorded in the faeces may be used to give a qualitative and possibly quantitative estimate of the natural diet of a herbivore. In a three-year study of the diet of Blackface hill sheep in Argyllshire, Martin (1964) showed that the pattern of grazing and selection of species by ewes was similar in each year (Fig. 8.2).

Red fescue (*Festuca rubra*) was the most frequent component in the diet of the sheep at all seasons, being especially sought after in March when fresh growth was appearing. There was a low intake of this species in May–June when the spring flush of other species occurred, notably *Molinia*, *Nardus*, *Carex* spp. and *Eriophorum*. Other species in the study area were selected at specific times of the year, *Calluna* becoming an important element in the diet from November–February, when other herbaceous material was unavailable, while *Juncus* spp. were eaten mainly in the October–March period. At all times of the year, Gramineae, Juncaceae and Cyperaceae formed the greater proportion of the diet, only becoming less attractive to sheep in the summer months when other herbs became available. The important general point derived from this study was that the maximum intake of a species occurred at the time of the year when the plant was most palatable, with secondary maxima occurring at other times when the animals were able to exercise little or no choice.

A study of the diet of Border–Leicester X Cheviot sheep grazing chalk grassland in Oxfordshire throws more light on grazing preferences. Over a period of two years, P. Hawes used the faecal analysis technique on sheep grazing at a density of 3 per a. (7·2/ha) for three months of the year in a series of paddocks on grassland containing more than 60 species of flowering plants. He found that the dominant grasses, *Festuca rubra* and *F. ovina*, contributed more than 80% to the sheep's diet irrespective of the season or time spent in the paddocks. Other grasses and sedges, notably *Helictotrichon pratense*, *Koeleria cristata*, *Briza media* and *Carex flacca*, made up most of the remaining diet. An interesting result from this study, which has implications in the management of chalk grassland, was the marked preference for *Carex flacca* during the winter months but this species contributed little to the diet during the summer. The faecal analysis technique provides only an estimate of the total diet as many herbs which were observed to be eaten by sheep did not appear in the faeces. Experimental studies showed that some soft-leaved dicotyledons, of which *Poterium sanguisorba* and *Lotus corniculatus* are good examples, were digested in the rumen of the sheep and did not appear as recognizable fragments in the faeces. Other herbs, such as *Helianthemum chamaecistus* and *Leontodon hispidus*, have silicified hairs which are not destroyed in the rumen and can therefore be identified.

The results from these and similar studies on selection by sheep enable the following general points to be made.

(i) They select leaf in preference to stem, and green (or young) material in preference to dry (or old) material.

(ii) Physiologically young material may be preferred because it is generally short or because it differs in chemical composition from old material.

(iii) Selected material is usually higher in nitrogen, phosphorus and gross energy than unselected food.

(iv) At high stocking densities the most abundant species in the grassland is likely to make the highest contribution to the diet.

(v) Species are not uniformly selected throughout the year, changes in preferences being dependent on the palatability of species at a particular time.

Of the many other factors which determine the selection of species in a pasture, *availability* is the most important. As the quantity of herbage on offer to the grazing animal decreases, selective preferences change and species which previously were rejected may be eaten. This knowledge may be profitably applied to a widespread management problem: the control of a highly productive, dominant grass such as *Zerna erecta*. This species is not eaten by sheep when grazing at low densities and other herbage is available but it becomes acceptable and is therefore 'controlled' when other herbage is scarce. This point was well illustrated in a sheep-grazing experiment on chalk grassland when *Brachypodium sylvaticum*, an unpalatable grass, was only consumed at the end of a long period of grazing in a paddock during which all other herbage had been eaten.

The presence of dead plant material (litter) in a pasture has an important effect on selection. As the proportion of dry or dead material in a pasture increases, the grazing animal is unable to eat an entirely green diet without decreasing its total intake. Under these circumstances there may be a complete suppression of preferences.

Arnold (1960) has drawn attention to the importance of grassland structure in influencing food selection. He showed that, in a pasture with a mass of tangled grass inflorescences in Australia, sheep were prevented from reaching the lower layers to graze green *Phalaris*.

Sheep and cattle move in a horizontal plane as they graze and select in a vertical plane (Arnold, 1960). Sheep usually eat the uppermost parts first, moving downwards, but rarely reduce a sward from 15 cm to 2 cm in one grazing. They prefer to move across the sward, gradually reducing the height, although the pattern of grazing is much influenced by the size of the area and the numbers of livestock.

On short, green pastures, sheep bite a number of leaves and part of the leaf is consumed. Cattle, on the other hand, curl the tongue around a tuft of vegetation, tearing the plant tissue, a method which seems to be less selective than that of sheep. Cattle-grazed pastures (Plate 25) are generally a mosaic of taller tufts interspersed with shorter vegetation. Horses obtain their food by close grazing and are reputed to be highly selective. They select the most palatable and nutritive species, often reducing them in quantity by over-grazing, while other areas in the pasture remain ungrazed and become coarse and rank. Although this effect of horse grazing is generally accepted observations at Willington, Bedfordshire have shown that when the stocking rate is high and herbage availability is low, the turf is kept uniformly short with no coarse vegetation. Under these conditions, horses have been observed to eat *Urtica dioica*!

Hunter (1964) has shown that home range behaviour of hill sheep is yet another important factor affecting vegetation. He demonstrated that a large flock of sheep divided into sub-flocks composed of families, the sub-flocks being restricted to parts of the pasture which differed in altitude, aspect and vegetation. The effect of changing the stocking rate cannot be accurately predicted by assuming that there will be a proportionate change over the pasture as a whole, since the site consisted of a series of home ranges. Most hill pastures vary greatly in quality, containing many different plant communities, some of which, such as the *Festuca/Agrostis* swards, are preferred by sheep while others are only grazed when no other herbage is available. Increasing or decreasing the number of sheep in sub-flocks may therefore alter the grazing pressure on certain plant communities because of changes in home range behaviour.

Effects of Dung and Urine

An important function of the grazing animal is the cycling of nutrients within the grassland ecosystem. The nutrient content of dung and urine and the volume returned to grassland depend on many factors of which the following are most important: type and age of livestock, quality of herbage being eaten, water content of herbage and weather conditions.

According to Herriott and Wells (1963), half-bred wetherhoggs, weighing between 80–120 lb (36–54 kg) return 500–2500 gm fresh weight of dung and 1·2–6·2 litres of urine per day per sheep to a pasture. During a normal grazing season they calculated that more than 12 000 gallons (54 552 litres) of urine and about 2 tons (2034 kg) of dung may be returned per a. (0·4 ha) under intensive sheep grazing. In a five-year study of sheep grazing on a grass/clover sward, Herriott and Wells (1963) calculated that the following amounts of nutrients were returned to the pasture per annum:

Mean lbs/a. (kg/ha) per year of N, P, K, Ca, Mg contained in dung and urine of sheep grazing a grass/clover sward (mean of 5 years' results)

N		P		K		Ca		Mg	
Dung	Urine	Dung	Urine	Dung	Urine	Dung	Urine	Dung	Urine
35·8	102·8	16·2	—	17·0	115	153	—	7	1·2
(40·1)	(115·2)	(18·2)	—	(19·1)	(128·9)	(171·5)	—	(7·8) R	(1·3)

It is clear from these data that most of the nitrogen and potassium returned to the pasture is present in the urine, while all of the phosphorus and calcium is present in the dung. The pattern of dunging and urination in a pasture may therefore have an important effect on the redistribution of nutrients within a grassland. Areas favoured as dunging sites often have high phosphate values which influences the vegetation growing there. The effect of sheep droppings on the yield and botanical composition of pastures has been examined by Sears and Newbold (1942) and more recently by Watkin (1954, 1957). They found that both dung and urine, applied separately, increased the productivity of the pasture, resulting in a clover-dominant sward with dung and a grass-dominant sward with urine. Indeed, Sears (1956) has suggested that, on a fertile pasture at Palmerston North, New Zealand, the annual return of nutrients in dung and urine was equivalent to 1 ton (1017 kg) of sulphate of ammonia, half ton (508 kg) of potash and 8 cwt (407 kg) of superphosphate.

Norman and Green (1958) studied the effects of cattle dung and urine on chalk grassland. They found that stock avoided areas that had received urine in the winter for 6–7 months, but when applied in spring, areas receiving urine were grazed within four weeks. Grassland treated with dung during the winter was avoided by cattle for 13–18 months but other workers have reported that dung patches are avoided for only four months. It is likely that availability of other herbage influences the time taken for cattle to return to these areas. Changes in floristic composition were small, although there were increases in *Festuca rubra* and *Dactylis glomerata* and decreases in *Chrysanthemum leucanthemum* in patches treated with dung and urine.

Both sheep and cattle dung are used as food material by fungi and coprophagous insects, the presence of dung increasing the range of insects found in grassland by providing another habitat. Dung deposited during the summer months is broken down more quickly than that dropped during the winter with the result that heavy winter grazing is likely to have a more deleterious effect on the sward on areas used as latrines. One way of overcoming this is by scattering the dung using a rake.

Effects of Treading

It is well-known that treading by livestock may have a considerable effect both on the structure and botanical composition of grassland, although there have been few experimental studies. The effects of treading on the soil (and ultimately on the vegetation) will depend on the nature of the soil and on soil moisture. Reduction in aeration, water penetration and regrowth have been recorded as a result of soil compaction. Thomas (1960) suggests that trampling by cattle is most dangerous in the tropics, causing erosion where the vegetation is worn away. Compaction caused by treading is not ameliorated in the tropics by frost heaving which occurs in temperate regions.

Sears (1956) estimates that a Jersey cow's hoof applies a pressure of about 45 lb/ in^2 (3·17 kg/cm) when walking, compared with 30 lb/ in^2 (2·1 kg/cm) by a sheep. On steep slopes cattle and sheep form paths which follow the contours of the slope and at high stocking densities this may result in erosion. Sheep also cause quite serious erosion by scraping out hollows for shelter on exposed hillsides, the scars of which are often visible many years after sheep have left the area.

On hillsides sheep and cattle tend to graze predominantly on the uppersides of paths and to defecate on the lower slopes. This leads to a zonation of vegetation away from the paths. Thomas (1959) showed that on the chalk the lower sides of paths were marked with an abundance of upright bromegrass (*Zerna erecta*) and tall oat-grass (*Helictotrichon pratense*), the line of tall grasses making the paths stand out when viewed from afar. Bates (1935) has commented that in upland pastures terraces formed by cattle and sheep are common. They usually consist of steps and 'risers', the steps having an abundance of perennial rye-grass (*Lolium perenne*) and white clover (*Trifolium repens*) in the outer margins, the centre of the track being bare.

The composition of the risers changes progressively, from clover, browntop (*Ag rostistenuis*), sweet vernal grass (*Anthoxanthum odoratum*) and York-shire fog (*Holcus lanatus*) to a vegetation with many herbs, the precise com-position depending on aspect and level of fertility.

On lowland grasslands in general, the effect of treading is most noticeable around gateways and water-troughs, where the vegetation is often eroded away leaving bare ground. Plants adapted to heavy trampling, such as *Polygonum aviculare*, *Coronopus squamatus* and *Matricaria matricarioides* quickly colonize these areas.

On wet alluvial marshes, the structure of the soil can be rapidly destroyed by excessive grazing and treading. This may also occur on heavy clay soils and leads to 'poaching',* a situation which invariably is associated with a loss of biological interest.

Systems of Grazing

The basic aim of the agriculturist is to obtain maximum productivity from an area of land at lowest cost, irrespective of whether the land is used for arable or grassland farming and often without considering the long-term effect of this management on what may be a non-renewable resource. The farmer operates within the constraints imposed by the fertility of his land, topography, climate and demand for particular crops. The influence of these factors is shown by the variety of grazing systems which have developed in a small country such as Britain. On the one hand we have the free-range system of grazing characteristic of our upland pastures and rough grazings, while on the other we have the intensive systems of management used on the highly fertile fattening pastures of parts of Leicestershire and the Romney marshes. Highly productive temporary grass is incorporated into the arable rotation in the well-known system of ley-farming.

Because every site is likely to differ in some feature from others, it is both difficult and dangerous to attempt to formulate management recommenda-tions for grassland in general. This reservation applies to the succeeding sections which examine, in more detail, systems of grazing which may be used by the conservationist for managing grasslands. These sections are based partly upon the results obtained from a limited number of grazing trials using sheep and cattle, and partly on observations and experience elsewhere.

* 'Poaching' is a term used to describe treading which breaks up the vegetation cover and churns up the ground surface.

Free-range grazing throughout the year

This system is widely used in the USA on the range-lands of the middle west and on the semi-arid grasslands of the south-east. In Britain and most of western Europe most montane grasslands and upland pastures are grazed in what may be described, broadly, as a free-range system. The pioneer studies of Boulet (1939) and Hunter (1962, 1964) on hill sheep in Wales and Scotland clearly show that home-range behaviour may often result in the majority of animals grazing only part of the total range available. Welch and Rawes (1966) have demonstrated in the Pennines that hill sheep graze heavily on the Agrostis–Festucetum association in preference to the Calluno–Eriophoretum, often preventing the flowering of many grasses and herbs. It is likely that under conditions of free range (Plate 26), particularly at low sheep densities, communities containing the more palatable species will be heavily grazed while other herbage will be avoided. This leads to patchiness in the structure of the vegetation, increasing the range of habitats for animal life.

At higher sheep densities, and particularly when the open range is divided into paddocks, the less palatable plants will be grazed; for example excessive grazing of Calluneto–Eriophoretum has resulted in a change to a Nardetum or Molinetum. For a more detailed discussion of grazing by hill sheep, the reader is referred to the valuable account by Rawes and Welch (1969).

Semi-natural grasslands in lowland Britain are rarely used as free-range pastures for sheep or cattle, partly because of the small areas available and partly because of the small size of the farms, compared with upland regions. Nevertheless, there are tracts of chalk grassland which could be managed in this way, although it is likely that this would lead to the gradual extension of less palatable species like *Brachypodium pinnatum*. However, in areas where this species is absent, free-range grazing would most likely lead to a mosaic of vegetation structures which is beneficial to nesting birds and the insect fauna.

Winter grazing

In temperate regions, vegetation makes little growth during winter and early spring with the result that grazing during this period has the effect of reducing the standing crop. Plant material produced during the summer months, which is not consumed, remains during the winter as dead or dying leaves and stems of low nutritive value (foggage). The accumulation of this material as litter often causes competition for light in the lower layers of vegetation. Winter grazing by cattle or sheep helps to break up this layer while the dropping of urine and dung hasten this process. Grant and Hunter (1968), in a study of the winter greenness of hill grass species, noted that by the middle of November, 75% of the leaves of *Holcus mollis* were dead or discoloured, *Agrostis tenuis* 50–75%, *Festuca rubra* and *Agrostis canina* 25–50%, while *F. ovina* and *Anthoxanthum odoratum* showed less than 10% discoloration. Provided stocking densities are kept high, the dead leaves of grasses will be eaten and an open, green pasture produced by spring. It may be necessary to supplement the diet of both sheep and cattle during the winter with concentrates or hay, and on areas where there is a high proportion of dead material, urea licks may help to provide the nitrogen necessary for efficient rumen activity.

A second important aspect of winter grazing is that many of the grassland dicotyledons are dormant during this period, usually with no growth above ground, and are therefore not directly affected by winter grazing. This suggests a way in which species of high competitive ability, for example many grasses, may be controlled without affecting the species which are dormant. Grazing in winter will undoubtedly lower the competitive ability of many grasses and will thus favour low-growing species the following spring. Calcareous grassland species, such as *Hippocrepis comosa*, *Anthyllis vulneraria*, *Anemone pulsatilla* and others, are known to benefit from winter grazing.

Many insects overwinter below ground or at the base of plants and at this time are not harmed by grazing. In areas where winter grazing removes an excessive accumulation of litter many plants are thereby able to flower the following spring and summer, and this is beneficial to many plant-feeding insects. However, hard winter grazing which removes all plant litter destroys the habitat of many invertebrate animals living in or on dead organic material and may be harmful to certain species of butterfly which pass the larval or pupal stage in the litter.

On some grasslands, for example washlands and alluvial meadows in the flood plains of low-lying rivers, winter grazing is impossible because of flooding. On heavy clay soils, poaching is more likely to occur as a result of over-grazing in winter and numbers of livestock need to be carefully controlled.

On hill lands, particularly at high altitudes, the ground is often covered by snow during the winter months and livestock are brought on to the lower grasslands in the vicinity of farmsteads. Thus winter grazing seems to be a practical proposition only in the lowlands and even there it may be necessary to supplement the diet of the animal with other food. Experiments on chalk grassland at Aston Rowant, Oxon, with Border–Leicester X Cheviot sheep have shown that sheep having a body weight of at least 100 lb (45·4 kg) or more are able to maintain condition during the winter months without supplementary food, except in the most severe winters. Fig. 8.3 derived from the Aston Rowant experiments illustrates changes in body weight of sheep which occur during the winter months when no supplementary feed is given.

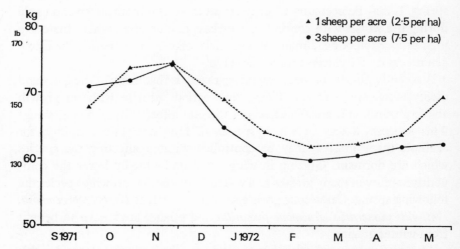

Fig. 8.3 Seasonal change in live-weight of Border–Leicester × Cheviot Sheep grazing chalk grassland at densities of 1 and 3 sheep per acre (2·5 and 7·5 sheep/ha).

At Woodwalton Fen National Nature Reserve, Hunts, Galloway cattle have been used since 1965, with good results, (Williams, Wells and Wells, 1974) to graze coarse fen grassland during the winter months although it has been necessary to feed hay and concentrates from December–March to supplement the low-nutrient content herbage.

Spring and summer grazing

Fig. 7.5 (p. 172) shows that grazing during the period April–September will affect the leaf and flower production and fruit setting of many grassland plants. The proportion of species which are able to flower and set seed depends primarily on the stocking density; at low levels many individuals of each species will be able to complete their reproductive cycle, while at high stocking rates very few will do so. However, a few species, such as *Thymus pulegioides* and *Origanum vulgare*, which are unpalatable, will flower and set seed even at high stocking densities. On the other hand, there is no evidence that grassland plants are adversely affected by grazing during the spring or summer, probably because most are perennials which are not dependent on establishment from seed for survival in the grassland.

Annuals, such as *Linum catharticum, Euphrasia nemorosa* and *Crepis capillaris*, increase in calcareous grasslands which are heavily grazed during the winter and spring period. The open conditions created by grazing, supplemented by the treading action of the animal, apparently create favourable conditions for their establishment.

Grazing at the time when the dominant grass is making its maximum growth (mid-May to mid-June) is an effective way of controlling competitive ability. This is most marked when the stocking density is kept high (24 sheep/ha on chalk grassland) for a short period, forcing the animal to eat vegetation which it would otherwise reject. However, even the more unpalatable grasses are acceptable at this early period when growth is rapid. Thus highly unpalatable species such as *Nardus stricta* and *Brachypodium pinnatum* are eaten in the April–May period but hardly at all at other times of the year. One of the primary objectives of management on most grassland nature reserves is the maintenance of floristic richness. In many cases, this can only be done by controlling the growth, development and spread of aggressive grass dominants and this is usually best achieved by grazing (or cutting) during the period late spring to early summer. Although this may prevent flowering in many species, some are able to produce a second crop of flowers late in the year following cessation of grazing. This is a convenient way of overcoming an unfavourable period.

Rotational grazing

Rotational grazing, in which parts of an area are grazed, while others are not (Fig. 8.4), is probably the best way of managing grassland for a diverse fauna and flora, unless there are species with specialized requirements (see also Chapters 5 and 11). The purpose of rotational grazing is to allow parts of the grassland to develop to different pre-determined stages. By grazing different parts of the area in succession, different structural formations are established, each with their own characteristic spectrum of plants and animals. At the same time, by regulating stocking rates and time of grazing, the more aggressive and dominant grasses may be kept in check

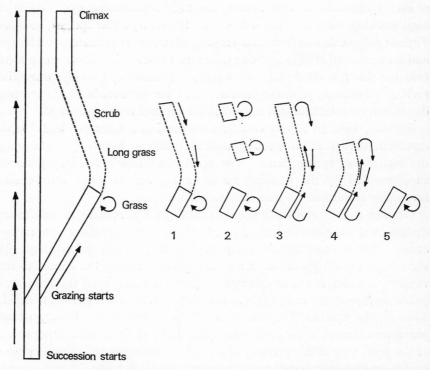

Fig. 8.4 Diagrammatic representation of natural and managed successions on grassland: 1. Reclamation of coarse grassland and scrub followed by annual management; 2. Theoretical management of short grass, long grass, and scrub; 3. Management of short grass, long grass, and scrub (long rotation): 4. Management of short and long grass (short rotation); 5. Annual management. (Arrows indicate direction of change.) (*From* Morris, 1971b.)

and prevented from overwhelming the lower-growing species. Longer rotational systems, for example every 5 or 10 years, are probably only suitable for grasslands overlying shallow soils where nutrient factors prevent excessive development of grasses and litter acumulation. In addition it will almost certainly be necessary to remove scrub during a long rotation, because grazing cannot control bushes once they reach more than 25 cm in height.

Short rotational systems, of 2–3 years, are likely to be the most effective way of managing grasslands on deeper, more fertile soils.

Rotational systems which incorporate grazing at different times of the year, concurrently with varying stocking rates, are likely to be the most flexible systems of management, but this method is unlikely to be practicable unless a large area of grassland is available and it is incorporated into some kind of farm system.

The most serious objections to a rotational system of grazing on areas managed particularly for wildlife conservation are practical and economic. For such systems to operate effectively, a permanent perimeter fence must be maintained together with internal temporary or movable fences to divide the area into paddocks. A water supply, feeding trough and access points are also necessary, the number and cost of which will increase with the complexity of the rotational system.

Despite these objections, it seems likely that as more and more grassland areas are managed for the conservation of wildlife, the value of rotational systems of grazing will be increasingly recognized (Morris, 1971b) and also the importance of different types of livestock, and other methods of management such as cutting.

9 Management and mowing

Cutting or mowing differs from grazing in three respects. First, it is non-selective, all vegetation above the level of the cutting blade being severed. Secondly, while dung and urine are returned to the sward by the grazing animal, cutting makes it possible to remove nutrients in the form of hay, or else return them in part, if the cut material is left as clippings. Thirdly, the cutting machine does not exert the same localized pressure on the ground as do the hooves of cattle and sheep. Fig. 9.1, taken from Norman (1960), is a much simplified attempt to show the relationship between competition and defoliation depending on whether management is by grazing or mowing. Competition between species in a diverse sward is a more complex process than that between a cereal crop and annual weeds. In the arable crop, both competitors normally grow to maturity and any reduction in yield of the crop is the outcome of a straightforward struggle for water, nutrients and light. In the sward, on the other hand, competition of this type only occurs when the grassland is neither grazed nor cut, but under management the balance between competitors is immediately altered. Numerous attempts to simulate grazing by using cutting techniques have provided information on the effect of cutting frequency on total yield and root-shoot ratios. On the other hand the effects on floristic composition and on the insect fauna have been little studied. Much of this work has been done in America in relation to the management of range grasslands (e.g. Laycock and Conrad, 1969; Owensby and Anderson, 1969; Everson, 1966) often with surprisingly conflicting results, particularly with regard to the effect of cutting on yield. Nevertheless, it is generally agreed that under field conditions the frequent, close removal of phytosynthetic tissue may decrease top growth of plants, and the more frequent and severe the defoliation, the greater is the reduction in dry-matter yield. Most writers are agreed that

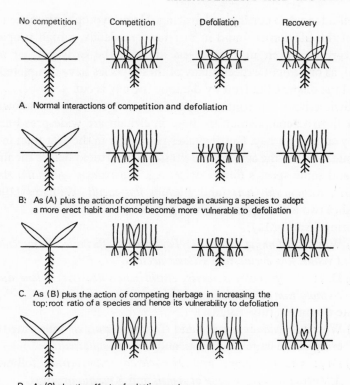

No competition   Competition   Defoliation   Recovery

A. Normal interactions of competition and defoliation

B: As (A) plus the action of competing herbage in causing a species to adopt a more erect habit and hence become more vulnerable to defoliation

C. As (B) plus the action of competing herbage in increasing the top:root ratio of a species and hence its vulnerability to defoliation

D. As (C) plus the effects of selective grazing

Fig. 9.1 Diagrammatic representation of the relationship between competition in a pasture under periodic defoliation. (*Redrawn from* Norman, 1960.)

defoliation is followed by a decrease of food reserves in roots and other storage organs, for example a fall in the levels of carbohydrates.

Mowing by hand is a traditional method on meadow grasslands in western Europe, particularly in upland areas where the steepness of the slope or the rocky nature of the terrain precludes the use of machinery. Hay-meadows are also found in the lowlands along river valleys where winter flooding and a high summer water-table make arable farming less attractive. Most of these grasslands are cut for hay from mid-June to late July, the precise time depending on weather conditions. On some of these grasslands a second crop is taken in late August, but more often the aftermath is grazed by cattle or sheep until the end of December, when the meadows are 'shut up' for hay. The outstanding feature of these grasslands is their floristic richness which exceeds that of any other type in Britain. Their floristic composition is primarily the result of long-continued management, natural selection operating in favour of those species which are well-adapted to cutting and grazing. Hemicryptophytes make up about 70% of the life-forms found in meadow grasslands, a high proportion of which produce their maximum leaf areas in the early part of the year. Indeed, as one might expect, many of these species have completed flowering and the dispersal of fruit by the time the hay is cut.

In Switzerland and southern Germany, wet fertilized meadows with a similar floristic composition to those in Britain are widespread and managed by cutting for hay. Cutting takes place later in the year than in Britain, but again, most of the agricultural grasses have fruited before the first cut is made and some species flower twice, e.g. *Taraxacum officinale, Anthriscus sylvestris, Ranunculus acris* and *Lychnis flos-cuculi*. Williams (1968) distinguishes two main aspects in these meadows:

(1) Spring (April–May).
   (a) Wet: *Cardamine pratensis, Taraxacum officinale, Ranunculus acris, Lychnis flos-cuculi,* and *Rumex acetosa*.
   (b) Dry: *T. officinale, R. acris, Anthriscus sylvestris, Silene dioica* and *Symphytum officinale*.
(2) Late Summer (July–August).
   (a) Wet: *Filipendula ulmaria* and *Cirsium oleraceum* following the flowering of the large *Carex* spp. and *Senecio aquaticus*.
   (b) Dry: *Crepis biennis* and *Heracleum sphondylium* following the flowering of *Tragopogon pratensis* and *Knautia arvensis*.

The growth of individual species of grasses in these meadows varies through the year. Growth is highest in the spring in *Anthoxanthum odoratum, Alopecurus pratensis, Festuca rubra* and *Carex panicea*, while in *Cynosurus cristatus, Bromus racemosus, Poa trivialis* and *Deschampsia cespitosa* early summer is the period when growth is highest. *Festuca pratensis, Poa pratensis* and *Holcus lanatus* develop later.

Fig. 9.2, which is taken from Traczyk (1968), shows changes in the standing crop and phenology of the more important species in the Strzeleckie Meadows, near Warsaw. It illustrates once again the general point that the reproductive processes in meadow plants are closely related to the type of management under which the assemblage of species has evolved.

The traditional management of meadow grasslands, by cutting in midsummer (Plate 27), would also seem to be best for nature conservation. There have been no ecological studies, as far as is known, on the insect fauna of meadows, but it seems reasonable to expect to find species with life cycles adapted to the longstanding management which these areas have received. It is possible, therefore, that rotational systems of management,

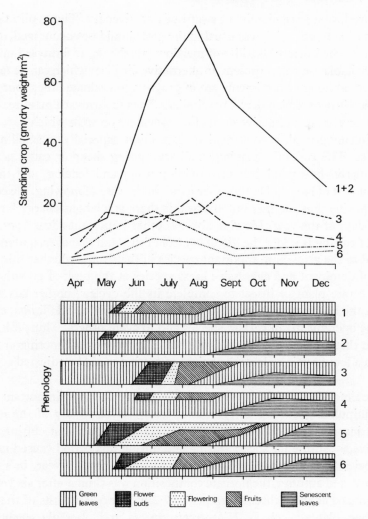

Fig. 9.2 Changes in standing crops and phenology of six important species from a Polish meadow. (*Redrawn from* Traczyk, 1968.) 1. *Carex nigra*; 2. *C. panicea*; 3. *Deschampsia cespitosa*; 4. *Festuca rubra*; 5. *Ranunculus acris*; 6. *Poa pratensis*.

which have been advocated for insect conservation on calcareous grasslands, are not necessary in meadow-land. A change in management away from traditional methods may also be deleterious to some birds which nest in meadow grasslands.

The application of inorganic nitrogen fertilizers to meadow grasslands has much the same effect as when applied to other types of grassland in that they produce a grass-dominant sward of low diversity. The application of farmyard manure which was usual in the past should be encouraged, as this maintains the inherent fertility of meadow grasslands. If farmyard manure is unavailable there is at present no alternative fertilizer which can be recommended, although experiments are in progress to evaluate other sources of organic nitrogen which may have similar effects to farmyard manure.

Mowing, as an alternative form of management on areas which have been traditionally grazed, is an attractive proposition, especially on small nature reserves. The cost of managing small areas using sheep or cattle is high, particularly the capital cost of providing water and fencing, and the recurrent cost of labour. On the other hand, in the case of mowing, the cost of the machine must be met but afterwards there is a labour charge for only a few days of the year. Mowing also has other important advantages. The time of cutting is under the direct control of the land manager, and may be related to the growth of important species in the sward, whether this be to control dominant and aggressive grass species at the peak of growth or to allow a rare plant or insect to complete its life cycle. Another advantage is that the height of cutting may be controlled within certain limits, enabling the operator to cut certain species and to avoid others. Thus, if it were known that pupae or larvae of a rare butterfly were overwintering or feeding on a low-growing plant, the height of cutting could be adjusted accordingly.

The time of year and frequency are likely to be the most important ways of manipulating cutting in grassland. In a six-year study of the effects of cutting on chalk grassland, Wells (1971) has shown that cutting is an effective substitute for sheep grazing and during this period caused no loss of species from the grassland. In plots cut three times a year, in spring, summer, and autumn, the floristic composition was similar after six years to plots grazed at about three sheep per a. (7·2/ha) for 6–9 months of the year. Prostrate plants, such as *Hippocrepis comosa* and *Asperula cynanchica*, increased under this treatment. Cutting once in spring was found to give a satisfactory control of the dominant grass *Zerna*, although cutting twice and three times a year was more effective in reducing its competitive ability. Defoliation acts differently on species according to the time of year when leaves and flowers are produced, fruit is set or the plant is dormant with no above-ground parts. In the experiment discussed above, although cutting at a certain time of year initially influenced the status of some species, this trend did not continue and after four years of cutting each species achieved a fairly constant level in the contribution it made to the vegetation in any plot. This result suggests that the effect on the total flora is to produce stability in the vegetation, which may differ in composition from that achieved with grazing but which nevertheless appears to maintain a similar degree of richness and diversity of species.

Cutting may have an important effect on insects which pass part of their life cycle within the inflorescences of plants. Langer (1959) showed that the date of cutting may have a significant influence on flower production in grasses. Working with single-spaced plants of timothy, he showed that cutting before 19th May (when the apical meristems of the grasses were below the level of cutting) had no effect on subsequent seed production but that cutting after 2nd June delayed flowering. Of particular importance was his finding that cutting after 28th July resulted in no inflorescence being produced in that year. For insects such as the frit fly, other stem-boring species, and members of the seed-feeding Rhyparochrominae, the effect of cutting, particularly in mid-summer, will be to remove a plant component essential for completion of the life-cycle.

The effect of cutting on the reproduction of the plants in the grassland community is not likely to be so important as for insects, because most are long-lived perennials which do not need to establish from seed each year. Nevertheless, it is important that they are able to produce seed occasionally and for the reasons discussed in Chapter 8 a rotational system of cutting is recommended on grassland sites where a grazing regime is traditional.

An important question for the conservationist is whether cut material should be left, or removed from the site. Two points need to be made: first, the physical effect of cut material on the vegetation beneath and secondly, the importance of returning or removing nutrients from the grassland. Both of these processes may have important effects on the composition of the grassland. On the more productive semi-natural grasslands, such as meadows, observation shows that if a swathe of cut grass is left on the ground, the underlying vegetation becomes yellow and will eventually die. Similarly if a 7 ft (2 m) high stand of gorse is cut and allowed to lie on grassland, the grass will probably die. In these conditions it is obviously best to remove the cut grass and use it as fodder. On grasslands with low productivity, which include most semi-natural grasslands in Britain (except for meadow grasslands), there is, as yet, no evidence that it is necessary to remove cut material, especially if a rotary cutting machine is used which shreds and distributes the cut material in a fairly divided form. In grasslands which have been unmanaged for several years so that an accumulation of plant material has occurred it may be necessary to remove this by burning before cutting is attempted.

Cutting and returning the plant material to the same area of grassland causes no change in the nutrient status of the sward, the nutrients in the grassland being re-cycled but not added to. In a replicated experiment on chalk grassland to test the effect on botanical composition of returning, or not returning, cut material, the mown vegetation was returned to the plots in a dried, finely divided form 2–3 days after cutting. After five years of these treatments there was no difference in botanical composition between the plots receiving nutrients and the plots from which cut material was removed. Both sets of plots had a rich floristic composition at the end of the experiment, in contrast with uncut plots in which the dominant grass, *Zerna*, formed more than 90% of the vegetation. In grasslands with low productivity, control of the dominant grass by defoliation is of far greater importance for management than the return, or non-return, of cut material.

Cutting has not been used extensively for the management of grassland chiefly because of the lack of suitable machines for dealing with conservation work. Gang-mowers pulled by a tractor can only be used on relatively flat surfaces such as golf courses and playing fields, although Chippendale and Merricks (1956) discuss the use of these as an aid to pasture management in the Romney Marshes. Reciprocating-blade machines used in haymaking are most suitable for use on flat land with a minimum of obstacles such as race-courses and are unlikely to be widely used except in meadow grasslands.

Allen scythes (Plate 28) which work on the same principle as the haycutter are useful for mowing grassland on flat areas, especially in places such as rides in woods or on grassy droves in fens, but they are unsuitable for use on rough ground.

Flail-type machines driven by the power take-off from a tractor are most suitable for use on rough grasslands situated on level ground or on slopes not exceeding 10°. On the steep slopes, which include many of our surviving semi-natural grasslands, rotary mowers which float on a cushion of air are most suitable because they can negotiate uneven ground. At the present time machines capable of cutting an acre (0·4 ha) per day of chalk grassland have been developed (Plate 29) and it is only a matter of time before advances in technology will make possible the development of larger, more mobile, machines.

10 The use of fire, fertilizers and herbicides in management

Grazing and mowing are the principal means by which grassland is managed for the purposes of nature conservation. Nevertheless there will be occasions when other methods are necessary either alone or in association with grazing or cutting. The purpose of this chapter is to examine the effect of these practices on the components of the grassland.

Burning

Fire is an important ecological factor in many of the world's grasslands, particularly in semi-arid regions where burning, annually or every few years, is a well-established practice (Campbell, 1960). Burning is carried out for one or more of the following purposes – to remove accumulated, inedible grass; to stimulate the growth of succulent and nutritious new grass; to influence the distribution of animals on the range; to remove a fire hazard, especially in areas which contain forests; to prepare a seedbed for natural or artificial seeding of desired forage species; to stimulate range grasses to produce seed and to encourage the growth of native legumes for forage and soil improvement. The ecological effects of fire are discussed in general terms by Daubenmire (1948) while Garren (1943) reviews the effects of fire on vegetation in the south-eastern United States. The numerous studies on the effects of fire on grasslands in the tropics and semi-tropics are detailed by West (1965).

In temperate regions, particularly in western Europe and Britain, studies on the ecological significance of fire have been mostly restricted to upland moorland communities dominated by dwarf shrubs, especially *Calluna* (Elliott, 1953; Whittaker, 1960; Gimingham, 1971). These communities have received considerable attention because of the importance of providing a succession of young *Calluna* for grouse and sheep (Watson, 1967). Taylor and Marks (1971) examined the effect of burning on the growth and development of *Rubus chamaemorus* in a *Calluna–Eriophorum* bog and concluded that rotational burning increased the vegetative spread and reproductive vigour of this dwarf shrub. The importance of fire as a factor affecting the dynamics of upland grassland communities in the Brecon Beacons in South Wales was realized by Crampton and Garrett Jones (1967), who drew up a list of plants according to their ability to resist fire. The species studied, in ascending order of resistance, were: *Deschampsia flexuosa*, *Calluna vulgaris*, *Festuca ovina*, *Nardus stricta*, *Vaccinium myrtillus*, *Agrostis tenuis*, *Molinia caerulea* and *Pteridium aquilinum*.

Burning is likely to be used most effectively by the conservationist in the reclamation of rough grassland which has been ungrazed or undergrazed and where there is a considerable accumulation of dead plant material (litter or mulch). In lowland grasslands in Britain, litter consists of dead grass, leaves and inflorescence stalks which have a high fibre and low nutrient content. This type of plant material is generally avoided by sheep and cattle when greener plant tissue is available. Furthermore, grass litter is only broken down slowly by micro-organisms and in under-utilized grasslands it tends to accumulate and 'smother' many low-growing species. This leads, eventually, to a species-poor sward which contains only coarse grasses and few herbs (Table 10.1).

Table 10.1 Mean percentage composition of ungrazed chalk grassland seven years after enclosure. Data from Knocking Hoe, Bedfordshire

	% (dry wt)
Litter (dead plant material)	73·7
Zerna erecta	18·6
Festuca ovina	0·3
Helianthemum chamaecistus	2·8
Poterium sanguisorba	1·5
Carex flacca	0·5
Filipendula vulgaris	0·5
Cirsium acaulon	0·4
Plantago lanceolata	0·4
Pimpinella saxifraga	0·3
Centaurea nigra	0·1
Other grasses and sedges	0·2
Other dicotyledons	0·3
Mosses	0·3

In the Cotswolds and the Peak District of Derbyshire the importance of regular burning of rough grassland to produce succulent new herbage has been appreciated for many years, considerable areas being burnt ('swaled') annually. Burning usually takes place in February or early March when the litter is dry (Plate 30) and before the growth of grassland species commences. The time of burning is important, most dormant grassland species being unaffected by fire, although above-ground organs may be destroyed. *Anemone pulsatilla*, a rare hemicryptophyte present in the Cotswold grasslands, is protected from fire by tough bud scales which surround the growing points (Wells and Barling, 1971). On the other hand, *Juniperus communis* is often destroyed by fire and only persists in those areas on shallow soils where litter production is low. The open, species-rich turf, found on many sites in the Cotswolds which have been 'swaled' for at least 50 years, is evidence that this is an effective way of managing grasslands dominated by *Brachypodium pinnatum* and *Zerna erecta*.

In a study of the effect of fire on limestone grassland in Derbyshire, Lloyd (1968) showed that while seedlings and young plants of *Crataegus monogyna* were destroyed, the older bushes regenerated from new shoots formed at the base of the plant. In the *Festuca–Helictotrichon* grassland studied by Lloyd, fire affected the dominant grasses *Festuca ovina* and *Helictotrichon pratense* in different ways. The more open tussocks of *F. ovina* were sometimes completely killed and as a result this grass never regained its position as a co-dominant, although it remained an important component of the vegetation.

An important effect of fire is the creation of small areas of bare ground between tussocks. These provide ideal conditions for the establishment of many grassland plants, both perennial and annual. Annuals such as *Linum catharticum* invariably increase during the first and second year following fires and subsequently decrease if fires are prevented.

Another effect of fire in grasslands, commented upon by many workers (e.g. Curtis and Partch, 1950; Lloyd, 1968), is the increase in inflorescence production in the year following a fire.

Few studies have been made on the effects of burning grassland on populations of invertebrate animals. Crawford and Harwood (1964) studied the effects of burning on lepidopteran larvae in Washington and recorded a fifteenfold reduction in numbers on burned plots compared with unburned ones, though burning did not invariably reduce the size of the population. It was suggested that most of the mortality was indirect, fire destroying litter in the grassland and thus exposing larvae to the rigours of overwintering under relatively open conditions. Tester and Marshall (1961) studied the effects of burning at two different times of the year and of grazing on numbers of Orthoptera and Coleoptera. Although they found that individuals of both orders were more numerous on plots with sparse litter than on treatments with deep litter, more precise data are required before the results can be accepted.

The effect of prairie burning on insect populations in Missouri is reported by Cancelado and Yonke (1970). They compared populations of some Hemiptera and Homoptera on prairie grassland, part of which was burnt in spring, with untreated areas near by. They found statistically significant differences between trap collections for Homoptera and Hemiptera in general and for the families Cicadellidae and Lygaeidae, with the greatest number of insects in each group coming from the burned area. The Miridae were also more numerous on the burned area, especially later in the season, and the authors suggest that near the middle of the growing season the ecology of the burned area offers conditions for a fast build-up of certain species of Miridae and other taxa, while later in the season there is a return to some condition of equilibrium. Unfortunately, no attempt is made to explain these differences in terms of the biology and ecological requirements of the species concerned. In New Zealand, burning is used extensively for controlling secondary scrub growth. Miller, Stout and Lee (1955) made a detailed study of the immediate effects of burning an 8–10 ft high (2·4–3 m) stand of *Ulex europaeus* and *Leptospermum scoparium* on the soil fauna and chemical status of the soil. The immediate effects of the fire were: (1) the destruction of surface litter killed many invertebrates and reduced the bacterial population. Partial sterilization occurred in litter that was moderately burnt and in the top-soil under large wood fires but much of the litter was relatively unaffected by the fire, (2) the vegetation was destroyed and its associated organic cycle was abruptly stopped, (3) large quantities of plant nutrients formerly immobilized in plants and litter were released to the soil in the ash. The amounts released were comparable with applications of fertilizers in farming practice.

Within a month of the fire, bacterial populations reached a peak and subsequently the microfauna, arthropods and earthworms increased in number, reaching a peak two-and-a-half months after the fire. At this stage typical litter animals were found in the subsoil, but as the food source decreased, the litter fauna also decreased. At the same time plant nutrients in the ash were being leached into the soil. With the establishment of grassland a new set of environmental conditions was provided for the soil fauna and microflora and Miller *et al.* showed that some of the original fauna disappeared, to be replaced by a different, but richer microfauna more adapted to grassland conditions.

The intensity, duration and time of year when burning takes place are likely to be important factors, but until more is known about the general biology and life cycles of insects and especially their overwintering sites, the effect of fire on them remains a matter for speculation. However, as a safe-guard against irreparable damage being done by burning, it is recom-mended that reclamation is done on a rotational basis.

Fertilizers

Farmers use fertilizers on grasslands for two purposes. First, to increase productivity in terms of fresh weight or dry weight yield per unit area, and secondly to manipulate the floristic composition so that the proportion of the more productive and palatable species is increased and that of so-called 'weeds' decreased (Heddle, 1967).

Inorganic nitrogen fertilizers are the most important chemicals used for increasing the productivity of temperate grasslands, applications of 300–400 units of nitrogen per annum per a. (0·4 ha) being applied to intensively managed grasslands in western Europe. Their use rapidly leads to a grass-dominant sward containing one to three species and so they are of little interest to the conservationist. The experiment reported by Smith, Elston and Bunting (1971), in which annual applications of nitrogen (N) and potassium (K) fertilizers were added to chalk grassland over a 10-year period, is worth examining in some detail as it accurately reflects the kind of change in floristic composition which has been reported in many other agricultural experiments with fertilizers, albeit usually over a shorter period. In their experiment, annual factorial applications of N and K fertilizers were made at rates of none or 47 lb N/a. (52·7 kg/ha) as nitrochalk (21% N), and K at none or 112 lb K/a. (125·5 kg/ha) as muriate of potash (49·8% K). The yields of herbage dry matter for four years (1960, 1962, 1965 and 1970) are shown in Fig. 10·1. They are typical of the responses obtained over the whole 10 years of the experiment – a small effect of K, a large effect of N and a large positive N + K interaction. Mean yields of the four treatments from 1960–70 are shown in Table 10.2.

The fertilizers had a very pronounced selective effect on the botanical composition of the plots. In both mown and unmown plots, those receiving N only became almost pure stands of *Festuca rubra*. Where N and K were added, the *F. rubra* became taller but large herbs such as *Cirsium vulgare*, *Senecio jacobaea*, *Taraxacum officinale*, *Tragopogon pratensis*, *Pastinaca sativa* and *Filipendula vulgaris* were able to survive. In plots receiving neither N nor K the number of species was larger and the turf resembled more typical chalk grassland (Fig. 10.2). With K alone, although the proportion of

Table 10.2   Mean annual herbage dry matter yields (above ground) 1960–1970 from plots receiving applications of nitrogen and potassium (see text for details)

(Kg/ha per year)

	No N	N	Effect of N
No K	730	1967	+1237ˣˣˣ
K	1013	2794	+1781ˣˣˣ
Effect of K	+ 283 (N.S.)	+ 827ˣˣˣ	
N + K interaction			+ 544ˣˣ

Significance: ˣˣP < 0·01; ˣˣˣP < 0·001.

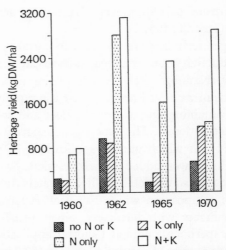

Fig. 10.1 Herbage yields (kg dry matter/ha) of chalk grassland 1960, 1962, 1965 and 1970 receiving the following fertilizers: None; 126 kg/ha of K; 53 kg/ha of N and N +K at rates given above. All fertilizers applied annually.

legumes increased, especially *Medicago lupulina* and *Lotus corniculatus*, it is interesting to note that *F. rubra* remained at about the same level as in the plots receiving no N. The results from this and many other experiments clearly show that inorganic nitrogen fertilizers should not be used on nature reserves or on other grasslands where the object of management is to maintain floristic richness and diversity.

Basic slag has been widely used as a source of phosphate and lime for about 100 years, primarily on hill pastures, to encourage clovers and, in-directly, to raise the productivity and feeding value of these pastures. Little is known concerning the long-term effect of this treatment on species composition, although the results from the Park Grass plots at Rothamsted

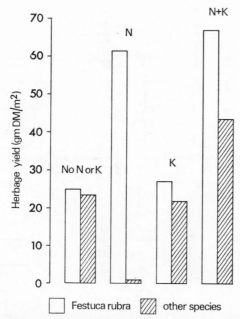

Fig. 10.2 Species composition (gm dry matter/m²) of chalk grassland, 1970. Left to right: no N or K, N, K, N + K.

(Brenchley, 1958) suggest that annual applications of other sources of phosphate (superphosphate) eventually lead to slight decreases in floristic richness. The effect is much less marked than with other fertilizers. For this reason, basic slag should not be regularly applied to grassland on nature reserves, although low rates of application at intervals of 5–10 years are unlikely to have a major effect on species composition. Further research is needed before this question can be satisfactorily answered.

The application of farmyard manure (f.y.m.) was the principal means of improving grassland before the manufacture of artificial fertilizers. Today, it is rarely used except on fields near to the farm in upland regions and on certain lowland meadow grasslands where traditional practices are still followed. Studies by Davies (1969) and Johnson and Meadowcroft (1968) on the effects, over seven years, of f.y.m. on meadow grasslands in the Pennines suggest that annual applications of up to 15 tons per a. (37 666 kg/ha) have little influence on the floristic composition of the grassland, the proportion of grasses to broad-leaved species remaining about the same. In addition the yield response was about the same as applying 40–60 units of inorganic nitrogen but without the adverse effects on floristic composition associated with this chemical. Moreover, there was sufficient available

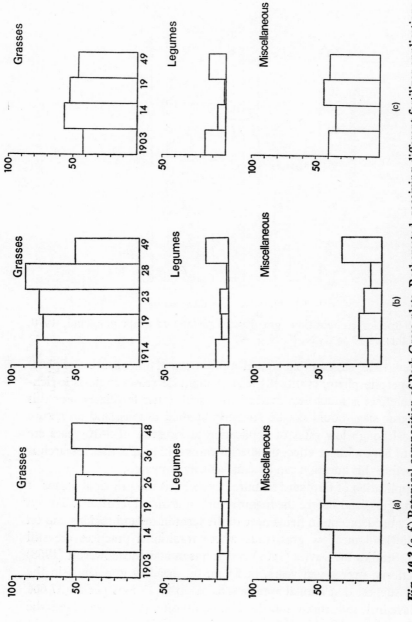

Fig. 10.3 (a–f) Botanical composition of Park Grass plots, Rothamsted, receiving different fertilizer applications. (*Derived from* Brenchley, 1958.) Vertical axes = percentage composition.

(a) Not manured since 1864.
(b) 14 tons/a (35 154 kg/ha) farmyard manure every fourth year since 1856.

204

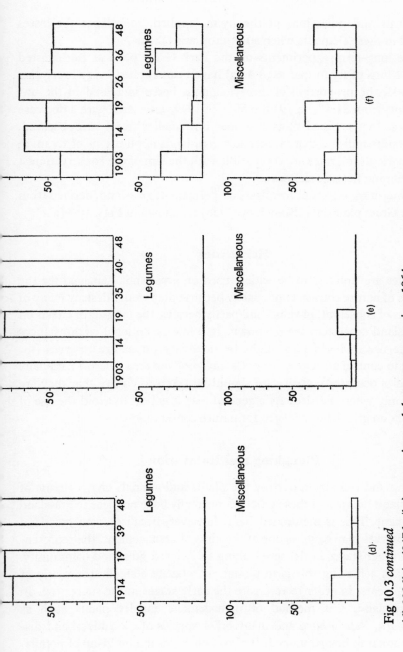

Fig 10.3 *continued*

(d) 220 lb/a. (247 kg/ha) ammonium sulphate every year since 1864.
(e) 440 lb/a. (493 kg/ha) ammonium sulphate; 392 lb/a. (439 kg/ha) superphosphate; 500 lb/a. (560 kg/ha) sulphate of potash; 100 lb/a. (112 kg/ha) sulphate of soda; 100 lb/a. (112 kg/ha) sulphate of magnesia. All applied annually since 1856.
(f) 392 lb/a. (439 kg/ha) superphosphate; 500 lb/a. (560 kg/ha) sulphate of potash; 100 lb/a. (112 kg/ha) sulphate of soda; 100 lb/a. (112 kg/ha) sulphate of magnesia.

205

phosphorus and potassium in the f.y.m. applied, to balance the losses incurred in these elements when a hay crop was taken.

In the long-term experiments on the Park Grass plots at Rothamsted where 14 tons of f.y.m. per a. (0·4 ha) have been applied every four years since 1904, the proportion of dicotyledonous herbs increased in the un-limed plots from 21·5% in 1914 to 50% in 1949, associated with a decrease in grasses. In the limed plots a similar but smaller trend was observed. These results indicate that on meadow grasslands applications of f.y.m. at normal agricultural rates are compatible with the aim of the conservationist in maintaining floristic richness.

The long-term effects of fertilizers on permanent grassland, derived from the Park Grass plots data (Brenchley, 1958) are shown in Figs. 10·3 (a–f).

Herbicides

Herbicides are unlikely to be widely used on grasslands managed for the purposes of nature conservation, partly because they would destroy many of the plants of conservation value, and partly because the long-term effects on the grassland ecosystem are unknown. In a few exceptional circumstances when the use of herbicides might be necessary on nature reserves (for example to control an aggressive grass 'invading' an area where a nationally rare species occurred), great care should be taken to ensure that they are applied only where needed. As a general rule it is best to avoid the use of herbicides on grassland managed for nature conservation.

Ploughing and Rotavation

Ploughing and rotavation destroy the plants and animals characteristic of the grassland biome and these practices normally have no place in grassland management. This is mentioned here, however, because some grassland nature reserves may have, as one of the aims of management, the represen-tation of other stages in the sere leading to a closed grassland community. One way of achieving this is to plough or rotavate a small area yearly or every 2–3 years, in order to re-create the early stages in the succession. In chalk grassland, this method of management enables plants such as *Iberis amara, Reseda lutea* and many other species of the ruderal and dis-turbed habitat to be maintained. It also encourages the build-up of popula-tions of phytophagous insects which are associated with arable weeds and so increasing the diversity of the reserve fauna. Longer-term rotations, for example, ploughing every 10–20 years, enables herb-rich communities containing unusual combinations of species to develop which are not usually found in older grassland. Nevertheless, it cannot be too strongly emphasized that ploughing or rotavation of grassland is generally totally alien to grassland conservation, except in the special circumstances described previously (p. 206).

11   The management of scrub

The management of scrub for nature conservation is a relatively recent concept and the detailed ecological research on which management should be based is almost entirely lacking. Scrub has to be managed in agriculture and forestry, and there is information about the planting of shrubs in hedges, shelter belts and in areas for game birds and animals. Cutting is carried out in hedges and coppicing cycles, and scrub is cleared from land which is to be afforested or farmed. Much of this information is useful for scrub management techniques for nature conservation. However, there are likely to be considerable advances in knowledge in the future, and it is important to keep records of the effects of scrub management on the fauna and flora of nature reserves.

Where there is a complex of scrub and grassland on a reserve first a broad decision must be made on the relative balance of these two types of vegetation that should be maintained. At the present time this is frequently weighted too heavily in favour of scrub, primarily because of financial and labour restrictions in setting up grazing and mowing schemes for grasslands. The increasing rarity of rich grassland means that these 'good' areas should have priority over most types of scrub. Where scrub is maintained in order to increase diversity on such grassland areas, it should be sited on the less floristically rich parts of the sites. If scrub is fruiting in areas adjacent to grassland, careful checks should be made to ensure that the management is efficient enough to control the establishment of woody plants.

Where the scrub is a climax community or succession is very slow, as on maritime or rocky areas, little or no management will be needed. If succession is more rapid then a dynamic system of rotational management can be used to try to provide for all the ephemeral species of plants and animals in the sere. A much longer time scale will be necessary, however, than the rotational systems that have been shown to be important in reconciling the conflicting needs of the fauna and flora in grassland (p. 188). Methods of attempting to maintain scrub as a plagioclimax are also possible, for example, the continuous removal of larger trees and bushes, and the maintenance of grassland glades and groups of trees with fringing scrub. It is thought, however, that the rotational methods will prove valuable in showing the relationships of the grassland scrub and woodland ecosystems, and so these methods have been emphasized in this account.

Rotational Management on Large Areas

Where it has been decided that scrub management is an important objective, and is to be related to the succession from grasslands, a rotational cycle can be set up. To do this, the succession from grassland to scrub should be allowed to take place. The scrub is then cleared, and the land brought back to grassland. After an interval under grass another period of succession through scrub follows. There should be a series of blocks of land at different stages of this rotational cycle.

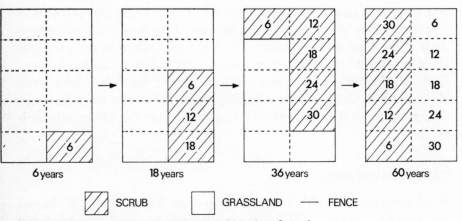

Fig. 11.1 Years of rotation and ages of blocks of scrub.

Fig. 11.1 illustrates these principles in a simple diagram where a cycle of 6 years has been chosen and where there are 10 blocks of land. Beginning from grassland the first block is allowed to progress to scrub for 6 years, and then the next block is taken into the rotation and so on until 30 years has elapsed. At this point the scrub is reclaimed back to grassland in the first block. The figure shows the different blocks of scrub after another 6 years when the rotation has operated for 36 years. At the end of 60 years all the blocks are in the rotation.

It is not possible to recommend a particular overall scheme for rotational management as every nature reserve will have to be carefully considered as an individual case, and the local conditions taken into account. Some of the factors that might need to be balanced in setting up rotational management schemes are discussed below.

Scrub succession time

This has not been studied in any detail but observations suggest that on deeper, richer soils grasslands may take about 15 years to develop to a fairly mature scrub, while on shallow, dry and nutrient-poor soils succession to this stage may take some 30 years or more. The presence of trees in the scrub is another factor which influences the succession. If there are many trees the succession is likely to pass quickly to woodland, but where there are very few trees the succession may be very slow indeed. A number of juniper stands on dry, poor soils for example have very few trees, and in southern England stands of old, dead and dying junipers may be some 80 years of age. These stands are often fairly even-aged with few young plants, and may be in danger of dying of old age (Ward, 1973). For this species therefore it may be necessary to consider very long-term rotation. An estimate of the likely time that succession might need to run on any particular site can often be made by ageing woody scrub plants in the general area. The rich flora of grassland and marginal species found in patchy scrub should not be shaded out by allowing succession to run on until closed canopy develops. The scrub should be cleared before this advanced stage is reached.

The grassland phase

It is known that eutrophication is likely to occur during scrub colonization and succession (Green, 1972). This is especially so where nitrogen-fixing species such as *Ulex* and *Hippophaë* are involved. If nitrogen or other nutrients have built up during the seral stages, removal of the scrub will result in colonization by tall species such as *Urtica dioica* and *Chamaenerion angustifolium* which grow on nutrient-rich soils. The general aim of grassland management is, therefore, to reduce the incidence of these tall coarse species, so that when the area is allowed to re-enter the scrub phase of the rotation, it will not be too difficult for the slower growing woody species to recolonize. Gay *et al.* (1968) recommend that it is important to remove all cut material from the area, including both the initial scrub and all the subsequent mowings, if the grassland is to be managed by cutting. Burning is also suggested by Green (1972). The swards which develop after scrub removal will not be as floristically rich as the more permanent types of grasslands, and may resemble ex-arable grassland (Cornish, 1954). Continuity of at least some of the grassland species during rotation should be ensured by the clearance of the scrub before complete closure of the canopy.

It is not possible to rule out management of seral scrub in relation to starting points other than grassland. Thus the grassland phase could be substituted by ploughing, and subsequently leaving the bare ground to be colonized by plants.

Management in relation to succession stage

There are few problems if the site is already grassland, but if it is partially or entirely scrub then some open grassland areas should be retained by thinning out trees and bushes in those blocks that will take longest to come into the rotation. Clearance work should be speeded up by extra expenditure in the early years.

The arrangement of blocks

The size and number of blocks in a rotation is likely to depend on a number of practical considerations such as cost of labour and machinery, as well as on the area and shape of the land available. If resources are limited the blocks should be small and the time interval between the different ages short. If, however, a larger area can be cleared at less frequent intervals, then fewer blocks might be preferable. The minimum requirements for the size and number of blocks are not really known. It is suggested that they should be $\frac{1}{2}$ ha or more, and that at least four age classes of scrub and grassland should be set up.

Landscaping on nature reserves is important. To preserve a pleasant visual impression the blocks should not be arranged in regular sequence providing this does not create problems for stock and machinery. Odd groups of trees or taller bushes should be left, to act as song posts for birds; Williamson and Williamson (1973) found that birds occurred in higher numbers where there were yew trees surrounded by scrub. The blocks can also be of irregular shapes, and be arranged to produce glades, and paths or rides with changing directions.

The costs of labour, materials and machinery for scrub clearance and management

Costs are important in determining the most practical systems of rotation and the number of blocks to be established.

Fencing will be an additional cost if stock are to be used in the management of the grassland areas. Temporary fencing may be suitable for shorter periods of rotation, while more permanent fencing may be needed for the longer periods. If rabbits are numerous, they should be excluded from the more recent scrub blocks to prevent grazing of young woody plants. Types of fencing used in forestry are discussed by Pepper and Tee (1972).

An Example of a Scrub Management by Rotation

No large scale rotational scheme for managing scrub is yet in operation on any nature reserve, although one has been suggested for Castor Hanglands National Nature Reserve. This scheme is discussed here as an example although modifications may be necessary once the rotation is set up.

Fig. 11.2a illustrates part of the Castor Hanglands National Nature Reserve where there is at the present time a good example of scrub with mixed southern shrubs, *Crataegus monogyna* and *Prunus spinosa* on Great Oolitic limestone. *Euonymus europaeus* is unusually common, while *Clematis vitalba* and *Viburnum opulus* are rare. There are a few scattered trees of *Quercus robur* and of *Fraxinus excelsior*. The scrub is of the 'thicket' type (p. 144), and has a closed canopy and few field-layer species. Its age is about 25 years on the east side but it is apparently progressively younger on the west. The grasslands (see map) include *Brachypodium pinnatum*, *Arrhenatherum elatius*, *Festuca ovina* and *Agrostis canina* as dominant species.

The proposed scheme for future management is shown on Fig. 11.2b. The approximate proportions of scrub and grassland already on the reserve are to be continued as it is important that the present grassland should not be disturbed. The rotation will therefore be confined to the present scrub area for which resources may be sufficient for clearance of ½ ha every 2 years. There is also the possibility of volunteers becoming available for more intensive work from time to time. The area has been divided into 12 blocks of ½ ha each including eight of scrub and four of grassland. The scrub phase will run for 16 years and the grassland phase for 8 years. The proposed order in which the blocks will be cut is shown on Fig. 11.2b beginning with the older scrub. Cutting might be speeded up in the early stages

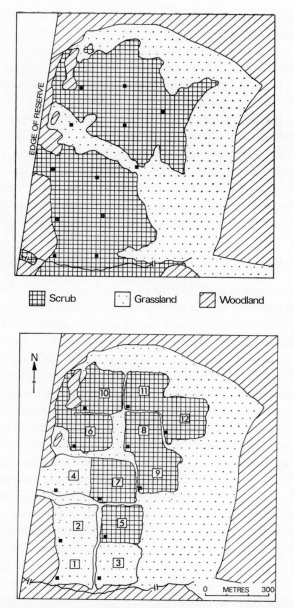

Scrub   Grassland   Woodland

Fig. 11.2 (a) Castor Hanglands National Nature Reserve showing the grassland and scrub on Ailsworth Heath in 1972. Markers in SW corners of each $\frac{1}{2}$ ha block. (b) The same area as it might be with rotational management of the scrub (the time scale is 16 years for the scrub succession, and 8 years for the grassland. The order of cutting for the scrub blocks is shown by the figures. See text p. 212 for further details).

and more than one block treated, because the scrub is already so advanced that grassland reclamation may be very difficult. Some blocks include managed grassland areas at the present time but will be allowed to revert to scrub in the near future. Landscaping is designed to conceal the straight edges of the blocks, which are not aligned so that glades will be created and the paths will not be straight. Existing woodland will be retained on the west side as a buffer zone between the reserve and the surrounding agricultural land, while scattered groups of trees will remain for birdlife and amenity purposes. It is not yet known how well the different scrub species will be conserved during the rotation, especially at the critical time of seedling establishment. The rarer and more interesting species such as *Euonymus*, *Viburnum opulus* and *Clematis* will not be removed indiscriminately during the clearance but will be left to act as seed-parents.

Management of the Species Composition of the Sere

Where scrub is managed by rotation as a seral community, the general aim is the maintenance of a rich assemblage of species of plants by natural colonization from seed-parents on the site. Many species of invertebrates with different ecological requirements will also invade the sere, so that within the series of treatments a wide range of animal life should be conserved. However, management for particular plants and animals might also be needed. This should be related to the autecology of the species but unfortunately such detailed information is not available for most scrub plants. The establishment of different species of woody plants will be influenced by factors such as the location of seed-parents, dispersal of the seeds, nutrient status of the soils, and the composition of the grassland. Grazing might also be used in some situations to decrease the proportions of the more palatable species. As so little is known about scrub management, the effects of any treatments on particular species should be recorded.

A management trial has been set up on chalk grassland to test the effects of (i) rotavation, (ii) heavy burning to kill the original vegetation, (iii) grazing and (iv) no treatment, on the course of succession to scrub. The full results of this trial are not yet available, but there are indications that the more open vegetation of the rotavated plots results in richer scrub communities. *Taxus baccata*, *Sorbus aria* and *Sambucus nigra* are commoner on the rotavated plots while *Crataegus* and *Betula* are favoured by burning. *Taxus* and *Sorbus* also occur on the burnt plots where competition from other vegetation has been reduced as in the rotavation treatment. *Taxus* does not occur on the untreated plots, but the large-seeded *Quercus* and

Fagus are commoner in the taller vegetation. Location is important: species such as *Rhamnus catharticus* and *Ligustrum vulgare* are only found on plots near to the main areas of seed-parents. Only a very few seedlings of *Juniperus* have been recorded, all on the plots nearest to the best seed sources.

Artificial seeding and planting may be necessary if more 'natural' management fails to conserve important species, or if the composition of scrub on a site is to be changed. In such cases seeds and cuttings from local sources should be used to maintain genetic continuity. Juniper is a species where this type of management may be needed. This is because the plant is becoming uncommon in some areas, and old stands are dying without replacement. This is likely to affect the fauna once the number of bushes begins to drop below 100. A quick method of establishment of young juniper is needed. This can be done by taking cuttings, rooting them in a mist propagator and planting them out. The construction of the M 40 motorway which crosses the Aston Rowant National Nature Reserve destroyed a juniper stand, but this is being replaced on the steep margins of the new motorway by the use of rooted juniper cuttings taken from the bushes on the original site. Seed may be sown in such situations, but much less is known about germination requirements for the best results and it is slower. Juniper seeds apparently require a double stratification treatment with an initial warm moist period followed by a cold moist stratification for several months (Livingstone, 1972). The seeds in these tests only germinated successfully in the field when they were buried. Old bushes are likely to yield a low proportion of viable seed, so berries from these sources should be carefully inspected before use. The later treatment of areas with juniper seedlings is not understood, but the reduction of competition from other plants might be important.

Other species of shrubs can be sown, or planted out, as necessary. Many woody plants, like juniper, require cold treatment to break their dormancy and a number are slow to germinate. The Woody Plant Seed Manual (Forest Service, 1948) contains useful information on many European species.

Once establishment has been achieved some interference in the natural course of the succession towards woodland might be necessary. Thus if trees are continuously removed the scrub formation will persist for a longer period.

Over-representation of some species can be changed by cutting, for example *Crataegus* or other species, to reduce interference with the slow-growing *Juniperus*. Juniper is often better managed as a pure stand in grazed grassland in the absence of other scrub species. Sheep-grazing appears to be better than cattle-grazing for this purpose as the latter often damage the bushes. *Buxus* too is often maintained in pure stands and in the past was coppiced on a rotational basis for the exploitation of its valuable timber. Elsewhere where its foliage is exploited for ornamental purposes *Buxus* stands may be interplanted with trees as the shading is thought to improve the colour of the leaves.

Another method of controlling succession by excluding forest trees has recently been described by Grime, Loach and Peckham (1971). This consists of placing fabrics with appropriate mesh sizes over the ground so that as the larger tree seeds are dispersed, they cannot reach the soil surface. The fabric can also be placed in or below the soil surface where it constricts the large roots or stems of invading tree seedlings without affecting the small plants. This dwarfs the trees and may eventually kill them by strangulation. Further research on these methods is in progress.

Scrub on Smaller Areas

Scrub can be very important on smaller areas to increase the diversity of habitats for wildlife. Bushes may be deliberately introduced or allowed to colonize natural areas set aside for this purpose. Several other uses which also conserve wildlife are listed briefly.

Landscaping

Shrubs are used to screen unsightly buildings or views, or can be grouped with trees to improve the general appearance of areas. Landscaping with trees and bushes is an important subject in relation to road and motorway verges, but other aspects such as safety and maintenance must also be taken into account (Way, 1969).

Hedges

Hedges form very useful boundaries which can be made stockproof and over a long period of time may be more economic to maintain than wire fencing (Hooper and Holdgate, 1968). If they are of the appropriate size and shape they form important refuges for wildlife on farmland. Moore (1968) has shown that 90 m of hedgerow will support one pair of birds, while Hooper (1968) records about 250 species of plants which occur frequently in hedges.

Shelterbelts and windbreaks

Many different shrubs and trees can be grown for these purposes taking into account the soil type and exposure of the district (Caborn, 1965) and the height and spacing required.

Game management

Many game animals, especially birds, find shelter and breeding sites in hedges, corner plantings, scrub woodland, and wide rides with scrub edges in forest areas. Several small coverts are generally better than one large one, the shrubs and trees should be arranged in layers for rising birds, and a margin provided for herbaceous plants and grass where game will find food and shelter (Coles, 1968).

Control of public pressure

Thorny scrub can form a good barrier to the movements of people, directing visitors away from more vulnerable areas of nature reserves, for example, where orchids are flowering. Both living and dead plants of *Hippophaë* have been found very useful for this purpose on sand dunes (Ranwell, 1972a).

Management of Small Areas of Scrub

The management of small areas of scrub for wildlife involves taking into account a few simple rules. Wherever possible, native shrubs should be planted or allowed to colonize naturally. If there are poor local sources of seed-parent species, it may be best to improve the richness of the natural communities by planting. If the woody species need to be cut to control their growth, it is best to do this on a rotational basis, cutting only part at any one time. Margins of grassland where the herbs and grass can be allowed to flower should be left around the scrub areas. Lastly, succession can be delayed if trees are regularly cut out.

Scrub Cutting and Removal

The clearance of scrub on wildlife areas must be carried out more carefully than in forests or on agricultural sites. Scrub should generally be cleared in autumn or winter in order to avoid disturbance to breeding birds. The surrounding vegetation must be carefully protected from herbicides if their use on bush growth is necessary, and damage to the terrain should be avoided. The cut scrub should be removed and burnt on bonfire sites, rather than allowed to rot down *in situ*, in order to reduce the levels of nutrients returned to the system.

Mechanical methods

Clearance carried out by hand labour is the most appropriate method in many areas especially where there are steep slopes (Plate 31), or where bushes and trees are very well grown, or set close in thicket scrub where machine access is difficult. Efficient hand labour can also be highly selective, making it possible to leave, untouched, species of special interest, and little damage to the terrain is done. Voluntary labour by organizations such as the British Trust for Conservation Volunteers is often available. If shrubs and trees are very young the individual bushes can be pulled up by hand. This method has been recommended for the control of *Hippophaë* on sand dune sites, especially those where invasion by the plant was not very extensive. Regrowth sometimes occurs, and treatment has to be maintained until the underground system expires (Ranwell, 1972a). For the larger woody plants, saws, secateurs, billhooks, axes, and lightweight mechanically-operated chain-saws can be used. Clearing saws, with a circular saw on one end of a shaft and a chain-saw motor on the other, are useful for stems of up to about 12 cm. They are held in a shoulder harness to provide balance, and can be used for cutting stems under bushes. The mechanical saws should be operated carefully and their indiscriminate use by inexperienced people who may carry out clearance work should not be allowed.

Various types of cutting or mowing machines are also available. Those operated and guided by one man have various types of rotating blades or reciprocating cutters. Most are only able to cut woody stems up to about 2 cm in diameter. Some are very effective for cutting regrowth in previously cleared areas and for the margins of scrub stands. More expensive machines which can be towed by tractors include hedge-cutting machines and the 'Bush-Hog' types of cutters. The cutting mechanism of the latter consists of rotating knives, suspended to allow a 360° rotation, and different hood sizes and blade lengths are available. Other types have rotatory chains on a central vertical shaft or rotating knives on a bar. These machines work best on stems of up to about 5 cm in diameter, and most give a rough cut which helps to prevent regrowth of the cut stumps. Some damage to the ground may be caused and a certain amount of regrowth from the cut stumps can be expected, but these machines may be useful for larger areas on reasonably level ground.

Mechanical uprooting can be done by winch and cable, or scrub removed with bulldozers and by using toothed 'grubber' blades. Remaining large stumps can be removed by the use of stump jacks. These methods are not normally recommended because of the damage to the soil and field-layer plants, particularly on clay soils and the need to follow-up with restoration work.

Ploughing is of limited application, but is used sometimes to prevent invasive suckering by shrubs into adjoining agricultural areas.

Herbicides

The subject of herbicides and woody weed control is only briefly considered in this book. Further useful information may be obtained from the Weed Control Handbook (Fryer and Makepeace, 1972) which covers all aspects of the use of herbicides including safety precautions and spraying equipment which are not discussed below. The control of woody species in forestry is described by Aldhous (1969).

Three herbicides are in general use for the control of woody species. These are 2,4-D, and 2,4,5-T and ammonium sulphamate. The first two are frequently compounded and sold as derivatives, often as the esters; and if emulsifiers are added they can be mixed with water or with oil. These two hormonal herbicides are sometimes applied as a mixture although 2,4,5-T is the most widely used. Ammonium sulphamate is more expensive than the other two compounds, and so tends to be used only where it is the most effective herbicide for the control of particular woody species. Recently, pichloram has been used, mostly on rangelands in America, Africa and Australia (Ivens, 1972), but it is a very persistent chemical and is unlikely to be suitable for use on nature reserves. All chemicals should be used very carefully on reserves because of the risks of damage to plants other than the target species.

Table 11.1 has been taken from Fryer and Makepeace (1972) and shows the susceptibility of some of the common woody weeds to the ester formulations of 2,4-D or 2,4,5-T, and ammonium sulphamate applied according to different techniques. Some species are resistant, or moderately resistant, to several of these chemical treatments, for example *Crataegus* spp., which nevertheless are susceptible to ammonium sulphamate. The advantages and disadvantages of the various methods of

Table 11.1 The susceptibility of common woody weeds to ester formulations of 2,4,5-T and 2,4-D and to ammonium sulphamate (*after* Fryer and Makepeace, 1972).

Recommended ranges of doses for esters of 2,4,5-T and 2,4-D:

Overall summer foliage sprays – 2·24 to 4·48 kg/ha in water.

Overall winter shoot sprays – 16·8 to 22·4 kg/ha in 1125 l of oil applied to run-off.

Basal bark and cut-stump treatments – 16·8 to 22·4 kg/ha in 1125 l of oil applied to run-off.

Ammonium sulphamate – 5 kg/5 l of water applied to run-off, or used in dry crystalline form.

The categories of response for these doses:

S – Consistently good control by suggested technique with little resprouting.

MS – Good control of aerial growth by suggested technique but generally requiring a higher dose or concentration than for 'S' category. Small plants killed; big ones may recover.

MR – Some useful effect from the higher dose or concentrations, but recovery usual.

R – No useful effect at the highest doses quoted.

Species	Diluent							Ammonium sulphamate to stump or stem. All seasons*	Remarks
	Water		Oil						
	Foliage		Dormant shoot	Frill girdle or basal bark		Stump			
	2,4-D	2,4,5-T	2,4,5-T	2,4-D	2,4,5-T	2,4-D	2,4,5-T		
Acer campestre (maple)	–	MS	S	MS	S	–	S	–	
Acer pseudoplatanus (sycamore)	–	MS	–	–	MS	–	MS	–	
Aesculus hippocastanum (horse chestnut)	–	MS	S	–	S	–	S	S	
Alnus spp. (alder)	MS	S	S	MR	MS	S	S	S	
Betula spp. (birch)	MS	S	S	MS	S	S	S	S	
Buxus sempervirens (box)	–	MS	S	–	S	–	–	–	
Calluna vulgaris (heather)	MS†	MR	–	–	–	–	–	–	† Apply June/Aug 3·4–5·6 kg in 670 l/ha.
Carpinus betulus (hornbeam)	–	MR	MS	–	MS	–	S	–	
Castanea sativa (spanish chestnut)	–	MS	S	MS	S	MS	S	S	
Clematis vitalba (traveller's joy)	–	MS	–	–	–	–	–	–	
Corylus avellana (hazel)	MR	MS	MS	MR	MR	MS	S	–	

Species							
Crataegus spp. (hawthorn)	R	MR	R	S	R	MR	–	S
Fagus sylvatica (beech)	–	MS	–	S	–	–	–	S
Fraxinus excelsior (ash)	MR	MR	R	MS	R	MR	MR	MS
Hedera helix (ivy)	R	MS MR	MR	S	MR	–	–	–
Ilex aquifolium (holly)	R	R	R	MR	–	–	–	MS
Juniperus communis (juniper)	R	R	–	–	–	–	MS	–
Ligustrum vulgare (privet)	–	MS	–	MS	–	–	MR	MS
Lonicera periclymenum (honeysuckle)	–	MR	–	–	–	MR	–	–
Pinus sylvestris (Scots pine)	R	MR	MR	–	MS	–	MR	–
Populus spp. (poplar)	MS	S	MS	S	S	S	–	S
Prunus spinosa (blackthorn)	–	S	R	MS	MS	–	–	S
Pyrus communis (wild pear)	–	MS	–	S	R	MS	–	S
Quercus spp. (oak)	R	MR	MS	MS	–	–	–	MS
Rhamnus catharticus (buckthorn)	–	MS	–	S	MS	–	–	MS
Rhododendron ponticum (rhododendron)	R	MS†	R	S†	R	MR†	R	MS†
Rosa spp. (briar)	MR	S§	–	S	–	–	–	S
Rubus spp. (blackberry, etc.)	MR	S	–	S	–	–	–	S
Salix spp. (willow)	S‖	S‖	S	S	–	MS	–	MS
Sambucus nigra (elder)	MS	MS	MS	S	–	S	S	S
Sarothamnus scoparius (broom)	S	S	S	S	–	S	–	S
Sorbus aucuparia (rowan)	MS	MS	MS	MS	–	MS	–	MS
Thelycrania sanguinea (dogwood)	–	S	–	MS	–	–	–	MS
Tilia spp. (lime)	–	MS	–	MS¶	–	–	–	–
Ulex spp. (gorse)	MR	S	MR	S	–	MR	MR	S
Ulmus spp. (elm)	MR	MS	MS	S	–	MS	MS	S
Vaccinium spp. (bilberry)	MS	–	–	–	–	–	–	–

† Use 16·8–22·4 kg in 112 l diluent
§ Good control at 1-12 to 2·24 kg/ha. Old canes more susceptible than young
‖ Good control at 1-12 to 2·24 kg/ha
¶ Applied in water

* All woody species listed here are susceptible to ammonium sulphamate. Where no symbol is given in the table, 2,4-D or 2,4,5-T is preferable because of lower cost. Where these materials are used, planting close to treated woody species does not have to be deferred for a period whereas in similar situations planting should be delayed by three months following treatment with ammonium sulphamate.

221

application of herbicides when in use on nature reserves are discussed below.

Stump treatment

The purpose of this treatment is to prevent regrowth from cut stumps, and it is considered to be one of the more useful techniques in conservation. 'Spot treatments' are widely used to prevent damage to near-by plants and animals. The problem of the later removal of unsightly dead trees and shrubs does not occur with stump treatment.

Woody species vary in their ability to produce regrowth from stumps of different sizes. *Hippophaë* for example is unlikely to regrow from larger stumps (Ranwell, 1972a). Other species, able to send out new shoots from larger stumps of over 25 cm, are not always successfully killed by stump painting, and in such cases frill-girdling of the tree before cutting may be more reliable (see below). Small stumps may be rather tedious to treat, but the method should control regrowth.

Stumps should be painted within 24 hours of cutting with solutions of 2,4-D or 2,4,5-T in oil, or can be treated with crystals of ammonium sulphamate. In the latter case the stumps are best cut in a 'V' shape to hold the crystals. The treatment should not be applied if rain is expected. Marker dyes, such as methyl violet can be used to keep a check on the stumps which have been painted.

Frill-girdling

This technique involves the painting of the herbicides into cuts made in the basal bark at about 80 cm height on trees and shrubs. It is more successful on larger stems and on those covered with moss than the basal bark sprays discussed below. The bark can be cut all around the stem or it can be notched using a small hatchet. The herbicide is applied in oil solutions by painting or by the use of a pressure oil can or water pistol. Ammonium sulphamate crystals can also be put in the notches. A tree-injection tool is commercially available consisting of a cutting blade used obliquely on the tree base, and a lever to release the solution of herbicide into the cut. Sufficient solution is put on to saturate the bark, and the treatment can be applied at any time from October to May, but it is most effective during January and March. This method avoids the hazards of spraying and drift, but it is only useful for older scrub with well-developed stems. However, there are sometimes difficulties in getting to the trunks because of low-growing branches. The other main problem with all methods of treating the trees and shrubs *in situ* is that after they are dead it is still necessary to remove them mechanically. Stands of dead woody plants are very unsightly and they continue to shade the species in the field layer.

Basal bark spraying

This method is very similar to frill-girdling, but is best applied to smaller trunks and stems up to about 12 cm in diameter. The herbicide is applied in oil to ensure penetration, and the bark is saturated at about 80 cm in height. The treatment is applied at the same time of year as the frill-girdling. The disadvantages may include access difficulties, the need to remove the dead plants, and risk of spray damage to near-by plants.

Foliar spraying

This method is not generally recommended because of the risk of spray drift damaging plants other than the target species. The hazards can be reduced to a certain extent by certain new developments in application techniques. These methods increase viscosity or change droplet sizes, the larger droplets being less likely to drift. The low-volatile esters of 2,4,5-T and 2,4-D such as 150-octyl can be used, as can the oil soluble amines in invert emulsions. The latter need special spraying equipment, however. Sodium alginate or hydroxyethyl cellulose can be added to water-borne herbicides to increase viscosity. Screening can also be used to protect plants not to be treated. Herbicide sprays should be applied during the growing season at the time of maximum foliage development. This method may sometimes be useful to prevent suckering regrowth on the margins of bushes, but more than one treatment may be necessary. It is less useful for thicket scrub as the spray does not always penetrate well enough to kill the plants.

Dormant shoot spraying

Dormant shoot spraying is a method used during the winter, but otherwise all the disadvantages of foliar spraying are involved. It has been used by foresters to control young birch.

Other methods

Burning is occasionally used in the control of species such as *Ulex*. It is not recommended, however, because of the danger to wildlife, and the possibility of the fire getting out of control. The bushes killed by the fire may also have to be removed. A licence is required from the County Agricultural Executive Committee for burning during the summer months.

12 The recording and measurement of management methods

In the management of grasslands for nature conservation emphasis has been laid on the need to formulate objectives of management. Once the objectives are clear a choice can be made of the appropriate methods and systems of management designed to achieve them. This process may be thought of as a continual refinement of both objectives and management policy.

A useful first step in the preparation of proposals for the management of a reserve is to write a *management plan*; published examples are available for the island of Rhum (Eggeling, 1964) and Aberlady Bay (Usher, 1973). Management plans normally consist of three sections; the *descriptive* part gives the location, size and boundaries of the reserve, together with an account of its wildlife interest and importance; the *objectives* list the objects of management; finally, the *prescriptive* part of the plan lays down the management procedure to be followed. For reserves which have been established for some time and in which management is almost continuous, successive revisions of the plan will be required. Past management, and its success or failure, will be recorded in the descriptive sections of these revised plans.

It is important that all management operations should be adequately recorded. A simple and effective method is to keep a management diary for a reserve; this is usually the duty of the warden. On small reserves the management diary will be combined with a general or phenological diary of events on the reserve, but on a larger reserve it is usually better to keep a separate management diary. The Nature Conservancy Council adopts a different method, more appropriate for an organization recording management on a large number of reserves. The 'event record' system uses specially designed cards on which 'events' including management are recorded.

225

These data are later transferred to 80 column punch cards. When using the management diary the vulnerability of a single unique record of management must be realized and it is often useful to have a periodic digest of work done which can be circulated, for example to a management committee. Whenever possible duplicate copies of all records relating to management should be kept.

Before specific management methods are applied it is important to record the state of the reserve. This is somewhat different from recording its scientific interest for the descriptive portion of a management plan. An account of the state of the reserve should emphasize those features which it is hoped to change by management. Thus if scrub is to be removed from a particular locality the size of the area should be recorded, together with a rough estimate of the species composition of the scrub. This may be helpful if at a later stage some predictions about invading shrub species are required. For all reserves it is essential to prepare maps, particularly of topographical features and of vegetation or plant communities. A management map, showing the areas to be treated or the expected state of the reserve after it has been managed, is also valuable.

Once management has begun there are many records which must be kept if the success or failure of the work is to be assessed. If areas of a reserve are grazed it is important to record the kind and breed of animal used. Even such a simple fact as whether sheep or cattle were used at a particular time may be difficult to ascertain after a period of years has passed. The breed, and the number of animals used, or some figure derived from it, such as stocking density, is of primary importance. Age, and any increase in the number of animals which may occur, should also be noted! Fences are a necessity when using grazing animals, so that a record of the type of fencing will be required. Annotations to maps should specify whether managed areas are enclosures, that is, pens in which the animals feed, or exclosures, which are areas from which they are excluded. Any changes in area of either enclosures or exclosures (for example by moving temporary fences) must be meticulously noted. The duration of the grazing period must be set down in detail. Management records of this kind are perhaps easier to keep when sheep or cattle are owned but if the animals belong to a local farmer, or licensee grazier, close liaison is needed to ensure that all relevant information finds its way into the management diary or other form of record keeping.

Although grazing animals are used by conservationists as a tool of management and not to make a profit, due attention must be given to animal health, for obvious reasons. Where possible, animals should be weighed before management commences and periodically during the grazing period. This is particularly desirable if animals are being used in winter without supplementary feed. Records of growth increment will be valuable in assessing management and regular weighing will enable a check to be kept on the general condition of the animals. Most conservationists will, from time to time, need the advice of a veterinary surgeon or know-ledgeable farmer. Although the making of a profit on grazing operations may not be obligatory it is often necessary to 'break even' and because many conservation organizations are not wealthy, strict costing of manage-ment is usually required, together with the drawing up of a budget. Common sense and a basic knowledge of financial matters may be all that is necessary.

The recording of mowing, burning or rotavating treatments is generally slightly less complicated than recording grazing. When grazing animals are used, fences form physical boundaries on the ground which can be readily checked against maps. Mowing and rotavating do not require the use of fences, so it is important to ensure that there are adequate markers on the ground which correspond appropriately with the management map. The type of mowing machine used should be recorded and also the height of cut, which varies according to the setting. Finally, the dates on which the vegetation was mown must be recorded so that the frequency of cutting can be determined.

Burning is a management activity which requires the greatest care and should normally be done only by experienced people. The direction and speed of wind should be carefully assessed and the burn should take place *against* the wind, not *with* it, if the operator is to have complete control. It is important that mistakes or accidental burns are recorded and not 'covered up', since it is obvious that management can be properly assessed only if correct records are kept.

Recording Change in Vegetation

The information which must be recorded in order to assess the success of management will depend on the objectives. In most cases the aim will be to record changes in the vegetation. Structural alterations are often brought about quickly by management but floristic changes are usually slower, although changes in the abundance of different species in the grassland may occur fairly quickly. Before embarking on a programme of recording it is well to consider the time and resources which are available to the reserve manager and his recorders. Simple methods are often better than more elaborate ones in this context. Ground photography gives a very useful pictorial impression of vegetation change, although trends may be difficult or impossible to measure by this method. It is usual and obviously most effective to take photographs from fixed points. Normally two markers are all that is necessary, one to fix the point from which the photograph is to be taken and one to indicate the direction. Successive photographs should be taken under the same conditions of weather and time of day, if possible. If photographs are taken at different times of year, this fact must be remembered during their interpretation. As with all photographs of this sort the view should include a standard indication of scale; a surveyor's pole is often chosen, or a man or woman of average height. Some fortunate managers of reserves may have access to aerial photographs showing the state of a reserve before and after (or during) management. Oblique or vertical air photographs can be of service, providing they are of sufficiently large a scale to show detail of management. Vertical cover has the advantage that it can readily be used to make maps of the area before and after treatment. Accurate mapping is a skilled process and best left to experts in photogrammetry, but rough maps can be made of stages in scrub management, for example.

In order to record vegetation change, suitable techniques must be used. Change in *structure* may be measured by simply placing a ruler vertically in the sward and reading off the height of the vegetation. It is usual to ignore inflorescences. Although a 'mean height of vegetation' is largely fictional very useful results can be achieved by this very simple method. Several measurements should be taken, at random, by the same person, and an average figure obtained. To record floristic changes in the vegetation following management various methods may be used to determine frequency or cover/abundance. Frequency is an estimate of the likelihood of a particular plant species occurring in a quadrat of predetermined size 'thrown' at random in an area of vegetation. To measure frequency a number of quadrats (preferably at least 20) should be placed at random, using a table of random numbers (obtainable for example, in Fisher and Yates, 1959) or a similar method. If this is not possible the quadrats should be placed by literally throwing the frame, or else a marker, to indicate where each quadrat is to be placed. The plant species occurring within each quadrat should then be listed in turn and percentage frequencies of occurrence worked out for each species. It should be remembered that particular measures of frequency are only comparable if they relate to quadrats of the same size. Large quadrats will be required for determining the frequency of shrub species compared with herbaceous plants. Further information on the use and limitations of this method can be found in Greig-Smith (1964).

Table 12.1   Frequency symbols and their meaning

d	dominant
co-d	co-dominant
a	abundant
f	frequent
occ	occasional
r	rare
vr	very rare
l	local, a suffix used to qualify other symbols

The abundance of plant species in a managed area of grassland may be estimated using the 'frequency symbols' familiar to all ecologists from the pages of Tansley (1953); these symbols are listed in Table 12.1. The recorder walks round the area and gives each plant species growing there a frequency symbol. It is supposed that each ecologist would give the same symbol if studying the same piece of vegetation. In recording management an assessment of the vegetation 'before and after treatment' will give some idea of the effectiveness of the work done. The method is highly subjective and it is surprising, as Greig-Smith (1964) says, that such good results have been achieved using it. The advantages of the method lie in its simplicity and quickness; the same recorder should assess the vegetation before and after treatment and the best comparative results are achieved if recording is done at the same time of year in each case.

Quadrats need not be used in assessing vegetation by means of frequency symbols, which record, rather imprecisely, the abundance of different plant species in the vegetation. It is often more helpful to record the *cover* of the species; quadrats must then be used. Cover is the area of ground which a species occupies, or the area of shadow cast by that species when illuminated directly from above. Cover may refer only to the top level of vegetation or be 'repetitive cover' and take into account the overlapping of different plants, and different parts of the same plant. Non-repetitive cover may be 'estimated' by eye and, indeed, forms the basis of two widely used methods, the Domin scale of cover/abundance (Table 12.2) and the Braun-Blanquet scale of cover (Table 12.3). The Domin scale is graded from 10 to 0, each point in the scale describing the cover or abundance of each species in the vegetation. For the higher points in the scale (4–10) cover is estimated; the lower points estimate the abundance of the species. The method is quick and although subjective is not open to all the objections which can be made to the use of frequency symbols. Cover/abundance may be measured or 'estimated' by eye. The Braun–Blanquet scale of cover is simpler than the Domin scale as it has only six, instead of eleven, points, and many ecologists prefer it for this reason. It is used in the same way as the Domin scale. Only a few quadrats need be placed in an area of vegetation for assessment using either scale. The quadrats may be *random* or permanent. For the assessment of change, following management, permanent quadrats are best. They are most conveniently marked out by placing wooden pegs at the corners; taller posts may be necessary in scrub vegetation. Random quadrats can be used in vegetation that is not uniform but a relatively large number must be placed to offset the variability between them. Quadrat size may be affected by theoretical as well as practical considerations. In scrub 10 m × 10 m is often a convenient size, whereas in grassland 1 m × 1 m is frequently used.

Table 12.2   The Domin scale of cover/abundance

Cover about 100%	10
Cover 75%	9
Cover 50–75%	8
Cover 33–50%	7
Cover 25–33%	6
Abundant, cover about 20%	5
Abundant, cover about 5%	4
Scattered, cover small	3
Very scattered, cover small	2
Scarce, cover small	1
Isolated, cover small	0 or X

Table 12.3   The Braun–Blanquet scale of cover

Cover 76–100%	5
51–75%	4
26–50%	3
6–25%	2
1–5%	1
<1%	+

By using the Domin or Braun–Blanquet scales, cover is 'estimated' or guessed, but it can actually be measured using the *point quadrat* method. This technique is an important tool for the professional ecologist but is rather too tedious to be used by any except the most enthusiastic conservationists to record change following management. The point quadrat frame is a relatively sophisticated piece of equipment (Plate 32), but is not expensive to construct. It consists essentially of a firm frame from which thin, pointed 'pins' can be lowered slowly into the vegetation. The plant species which the point of each pin touches first (for non-repetitive cover), or all the 'hits' the point of the pin makes before reaching the ground (for repetitive cover), are scored. The number of hits on each plant species out of a hundred total hits is the percentage cover of that species. Greig-Smith (1964) discusses some of the theoretical aspects of measuring cover using point quadrats.

It often happens that an area of managed grassland changes slowly in floristic composition from one part to another, as when soil becomes progressively shallower in passing from the bottom to the top of a downland slope. Management of such areas cannot be satisfactorily assessed using a few quadrats but may be recorded using one or two *transects* across the area of floristic change. A string stretched across the area to be recorded will serve as a *line transect*. It is sometimes possible to measure the length of line which each species in the vegetation touches. The proportions of the total length of transect which this represents is a measure of the abundance of that particular species. More usually, abundance is better measured using a quadrat. Line transects are also used to place quadrats or point quadrat frames at set distances along a gradient. Belt transects (narrow belts of vegetation of constant width) may be used for the same purposes. It will be seldom necessary to record abundance or cover along a transect line except at a very few (perhaps three or four) points, unless recording management change is being combined with other studies on the vegetation.

Very often the conservationist is less concerned with general changes in the vegetation following management than in the effects of treatment on particular species. So far in this chapter it has been assumed that a competent field botanist is available to record the vegetation and identify all the grassland plants in their various growth stages. If such a paragon is not available the recording of management effects on a particular plant species, or group of species, may be better than not recording the effects of management at all. In many grassland reserves there are particular plant species (for example orchids, fritillary, pasque flower) whose response to management must be ascertained. Plants which flower are both easier to count and record, and a better indication of the success of management than vegetative plants. If a species is long-lived and distinctive, as are many orchid species, it may be possible to map all the plants and record flowering success under management. A rapid method for determining the position of each old plant is to use two tape measures or marked strings, from two set points, as co-ordinates. If random quadrats, or sample areas, are to be placed within an area of vegetation, to record flowering success of a particular species, it will be necessary to obtain statistical advice on the type of distribution shown by the flowers of the species. Permanent quadrats or 'sample' areas can be used to give comparisons between years but are likely to be most unreliable if an estimate or the total number of flowers produced is the objective. The numbers of flowers or inflorescences produced by a species under management is only one measure of the success, or performance, of that species. In some cases it is more important to know if viable seed is set and, if so, how much; in others the actual number of plants growing in an area may be important, whether or not flowers are produced.

Agricultural botanists and production ecologists are often interested to know how much vegetation is present at a given time in a particular area (the standing crop) or to obtain estimates of production or productivity (Chapter 3). It is unlikely that conservationists are likely to require such information except in the course of specific research. Methods of estimating production and productivity are given in such textbooks as Milner and Hughes (1968).

Recording Change in the Fauna

On those reserves where conservation of animals is an objective of management, and particularly where rare or uncommon animals are known to inhabit managed areas of grassland, it may be necessary to record the effects of management on the fauna. Many conservationists do not have the time or expertise to record the effects of management on animals and because vegetational changes under management are both more obvious and more easily recorded than changes in the numbers of animals, it is often tempting to try to deduce the effects of management on animals from those on vegetation. While such an approach is better than none it can often be misleading. On the whole, however, there is a good correlation between *structure*, in both individual plant species and the gross vegetation, and the numbers of animal species and individuals inhabiting a grassland. Patchiness, or an intimate mixture of short grass and tussocks of taller grassland vegetation, is often a good sign of general faunistic richness.

Recognizing the difficulties of recording a very large number of animals and their habitats, Elton and Miller (1954) proposed a scheme of recording of 'habitat types' as a simple method of assigning animals to one or other of a relatively small number of categories. The scheme has been discussed more fully by Elton (1966). A modification of this classification is in use for site description and comparison (SPNR, 1969) but both schemes have, in general, too coarse a 'mesh' to be useful in recording the effects of management. An exception might be 'reclamation management', for instance, reclaiming scrub to grassland.

Just as the standing crop of vegetation can be recorded so can the standing crop of fauna, or at least of invertebrates. According to the techniques used, insects and other animals in standard samples may be counted or their biomass ascertained. These figures can form the basis of comparison between treated and untreated areas. However, such methods are of obviously limited appeal and importance to conservationists and managers of land; they are mentioned here because of their relatively small involvement with taxonomy; specimens need be identified only to order, or in extreme cases may be lumped all together.

In most cases where the broad effects of management on the fauna are to be studied recourse will be made either to an indicator species or a group of species which respond to management. An indicator species or group may be envisaged as being representative of the whole fauna although generally this concept is an over-simplification unless the indications looked for are clearly restricted. Often a species is chosen as an indicator because it is, in fact, one of the most important, or the most important, species present in an area. It is not really correct to think of these as indicator species at all, as they are not necessarily representative of the whole fauna but are important in themselves. Species in this category will usually be the more obvious, larger and attractive ones, such as birds, mammals, butterflies and perhaps grasshoppers and bumblebees.

Accurate *rapid* survey of animals in these or other groups can be beset with difficulties and inaccuracies, but is often the only form of comparison possible. The aim in most cases will be to *compare* the occurrence of the particular species or group of species either in areas under different forms of management, or in the same area before and after treatment, often in successive years. For the results of such comparisons to be at all meaningful it is important to standardize recording. The most useful way to record is likely to be by comparing numbers taken per unit effort. Thus the number of grasshoppers collected per hour or the number of butterflies of one species seen per twenty-minute period are valid figures for comparison, provided the numbers are collected or observed under the same conditions of weather, time of year and day, and preferably by the same recorder or by a group of recorders who have standardized their methods of collecting or even of observation.

Methods for estimating numbers of small mammals using the Longworth trap are given by Southern (1964) who points out that the interpretation of data from live trapping may be affected by several factors, particularly learning behaviour. In many lowland grasslands, the skylark and one or two other species may be the only common birds but other areas such as the grass heath of Breckland and flood-meadows and washlands may have a bird fauna of considerable interest (Chapter 5). On types of scrub there may be high numbers of breeding passerine birds (Chapters 5 and 6). Methods for censusing birds have been evolved by the British Trust for Ornithology in their 'common bird censuses'. A variant of the 'numbers per unit effort' is the line transect, which has been used especially for recording birds. A fixed transect line is walked by the recorder and the numbers of animals of the species to be studied is noted, either by visual observation or, in this case of birds and possibly extendable to grasshoppers, by hearing their song.

'Numbers per unit effort' is the simplest method of comparing one area with another or the same area in different years; it is really a method of quantifying visual comparisons of the type: there are more of x in this area (year) than in the other one. More meaningful are sampling methods which compare the densities of, usually, a given stage in the life-history of a species in different areas or different years. For sessile or sedentary animals ordinary quadrats may be used to estimate density in much the same way as numbers of individual plants can be estimated. The same difficulties apply, too, except that individual animals are usually more easily distinguishable compared with some plants. Random quadrats may give good comparisons of mean density of a species but the estimation of total numbers will depend on the way in which the species is spatially distributed. If the numbers of animals per quadrat are very variable, statistical advice should be sought. Although most animals are active rather than sessile or sedentary, quadrats can be used with some success to estimate mean density of some quite active animals such as grasshoppers. It may be important to have a number of recorders, or helpers, to catch the grasshoppers as they jump out of the quadrat following its positioning in the vegetation. Even with very active animals such as butterflies it may be possible to estimate density using quadrats, by working with the early stages or by counting the adults at dusk or earlier, since butterflies rarely fly, except when disturbed, after 18.00 G.M.T. Generally speaking, however, butterfly caterpillars and pupae are very difficult to find and adult butterflies have a very definite 'roosting behaviour'. Thus most of the 'blues' pass the night on the stems of grasses or other tall vegetation, although the females may oviposit and the larvae feed on much shorter plants. Recording the butterflies while 'roosting' will normally give a bias towards tall vegetation.

One species of Lepidoptera which is not a butterfly is particularly useful as an indicator species because of its pupating behaviour. The common six-spot burnet, *Zygaena filipendulae*, spins a cocoon high up on a grass stem. These are readily counted and estimates of cocoon density can be made relatively easily. Moreover, it would seem that *Z. filipendulae* can only inhabit fairly tall grassland, not being found on recently grazed or managed swards; however, the evidence for this is observational and no detailed study of the species' ecology has been made.

Various 'standard' methods can be used to compare one area with another, or the same area at different times, even though they are not strictly sampling methods. Thus on grassland the sweep net may be used to record numbers of different insects by taking an equal number of 'sweeps' on the two areas or at different times. In scrub the beating tray has a similar function. Weather affects the catch of some species (Hughes, 1955) and probably of most. Many forms of trapping for invertebrates, on the other hand, are open to objections. These include light-traps, pitfall traps, water or window traps, sticky traps and aerial traps. Light-traps are widely used to collect moths and some other insects but it is very difficult to standardize conditions, and almost impossible to do so if trapping is done on different nights. Perhaps least objectionable is the comparison of different areas by trapping on the same night, or series of nights. But even in this case the proximity of other biotopes and other uncontrollable factors are likely to make the results of doubtful validity. A possible method of using light-trapping to compare different areas, or the same area under different management regimes, is discussed later in this chapter.

Pitfall trapping, for beetles, spiders and other invertebrate animals, is equally unreliable for comparisons between treated areas, either in time or space. The only real value of pitfall trapping is to record what species of active invertebrates are present in the area. If there is no, or little, differential behaviour of activity in time a phenological record may be obtained in some circumstances. Differences between numbers of one species and another in pitfall traps may have no relation to the true densities of the two species. On a grassland site in Norfolk two millipedes were found in pitfall traps, one, *Schizophyllum sabulosum* commonly, the other, *Cylindroiulus latestriatus*, much more rarely. When the two species were *sampled* the mean densities recorded per square metre were $2 \cdot 3 \pm 1 \cdot 5$ and 173 ± 57 respectively! (See also Sutton, 1972, p. 106). The comparison of different areas using pitfall traps is difficult and any quantitative comparison of areas managed in different ways quite impossible. Greenslade (1964) has commented that the only valid result that can be obtained is if numbers of a given species in pitfall traps are *greater* on the denser, more obstructed, biotope. In the grassland context this means that a larger number of animals recorded from unmanaged grassland compared with managed is likely to be a true indication of the actual situation. No valid conclusions can be drawn from any other kind of result.

Capture–mark–recapture, or Lincoln Index, methods of estimating numbers can be used to compare different areas in favourable circumstances, or, more easily, to compare the same area in different years. In essence the method is simple: a number of animals (x) is collected, marked in a distinctive way, and released into the population. Subsequently a further collection is made (y) of which some (z) will be marked. The population may then be estimated as xy/z. Some large assumptions are made in this simple exposition, however. It is assumed that there is no mortality or natality (birth) in the population and that the x marked animals are dispersed randomly in the population following marking and before the second collection, y, is made. Dispersal of animals into or out of the study area is ignored. A detailed examination of capture–mark–recapture methods is outside the scope of this book; a concise account is given in Southwood (1966 Ch. 3). Despite the difficulties and the labour in calculating population numbers from capture–mark–recapture data, the method is popular among amateur naturalists as well as being widely used by professional ecologists. The limitations of the simple methods of analysing the data must be borne in mind, however. It is probably true that most sets of data, at least where marking is done over a reasonably long period and where the marked animals are well mixed in the population, can be analysed at different levels of sophistication and accuracy. Thus, providing a sound method is adopted in the field, data may be subsequently examined more deeply if the opportunity arises.

All the methods of counting, or estimating, actual numbers or densities of animals which have been considered relate to the situation at one point of time. It must be accepted, however, that simple estimates of population numbers or density can be misleading in that such information cannot indicate whether the population is increasing or decreasing in numbers, nor can it give any indication of the rate of increase or decrease. Even successive yearly estimates may not yield satisfactory evidence of the success or failure of management because of the possibility of fluctuations in numbers due to factors other than management. Butterflies, in particular, among grassland animals likely to be studied, fluctuate in numbers from year to year, often in relation to weather. On the other hand, some responses of animals to management, or the cessation of management, are so marked that they tend to mask any year-to-year fluctuations. Because of these factors the presence of a given species as a breeding population in a managed area of grassland may, in some cases, of itself be taken as an indication of the success of management.

The most satisfactory method of assessing the effect of management on any particular species is, in theory, a study of the population dynamics of that species in relation to management. Population dynamics, however, as a field science, is very much a matter for the professional ecologist. Such studies demand not only considerable expertise in ecological and statistical techniques but also the expenditure of much time and effort in the field and laboratory. They are unlikely, therefore, to be used in assessing the effectiveness of management, and recourse must be made to simpler techniques, the results of which may be less certain in indicating trends and definite effects.

Species Diversity and Management

Ecologists interested in the associations or communities of animals in grassland, or any other kind of biotope, often seek to define the characteristics of such 'many-species populations'. We have seen that species-richness, the number of species occurring or breeding in a managed area, is one criterion of management success. Diversity is a characteristic of faunas which is an elaboration of richness. The relevance of diversity to nature conservation has been briefly discussed by Morris (1971a). Although diversity studies are unlikely to be of use in assessing management success it is worthwhile mentioning that there is a property of one particular index of diversity which is of at least theoretical importance in this context. The index is Williams' α (Williams, 1964), given by the formula

$$S = \alpha \log_e\left(1+\frac{N}{\alpha}\right)$$

where S is the number of species occurring and N the number of individuals. α varies less with sample (or collection) size than do all other indices of diversity. In theory, then, any reasonably large collection of, say, moths collected at a light-trap should give a value of α which can be compared with α for a collection made at a contrasting site or at the same site after or during management. Each collection of moths, however, must conform to a particular distribution of individuals among species (the logistic or log-normal). At the moment this method of assessing the success or failure of management seems to be rather a 'long shot'. It is being used in very extensive studies in an attempt to assess land use changes, among other aspects, by Taylor (1968). The possibilities might appeal to the many entomologists who run light-traps, providing methods were standardized.

Monitoring Long-term Changes

In many nature reserves management may be impossible, undesirable or intermittent. In such areas it is often desirable to monitor the changes occurring over a relatively long period. For such a purpose it is often best to use the simplest and most easily reproducible techniques available. Photographic records (see p. 228) may be of particular importance. Often records need be taken relatively infrequently, but regularly; once every five years may be reasonable on many reserves. Such records will aim to assess changes attributable to succession rather than to management. Thus maps showing the extent of scrub may be important and annual counts or estimates of the numbers of a rare or valuable species be necessary so that the species is not lost through lack of management. It is often the case that the rate of change is as important as the change itself. Thus the advance of scrub into a grassland reserve may be acceptable provided the rate of advance is not too great, or the continuing reduction of an orchid population be tolerable provided that the rate is slow.

With changing personnel and staff and circumstances, as well as with the improvement of techniques, no scheme of recording can last for a long period, let alone for ever. Simple recording done regularly and completely is nearly always better than more sophisticated work performed irregularly or only in part. Hence self-discipline and discipline of others is important here. It is a good idea to arrange the work of recording, if done each year, well in advance and tell participants before they plan their summer holidays or take on other work. In situations where work is done voluntarily, much may be achieved by encouraging and maintaining individual enthusiasm, but it is not always safe to rely on it once the novelty of the work has worn off.

Sometimes (only occasionally one hopes) it will be necessary to make a radical and rapid change in management as a result of recording its effects. Often this will be because of the discovery of the existence of a rare or endangered species which could suffer as a result of management. More usually, experience will show that some modifications of management are necessary in order to achieve the objective of management which is the first priority. One must remember that the management plan only serves as a guide to achieve the objectives. Just as the plan is useless if prescribed management is not done, so a slavish adherence to the plan which is not achieving its objects is worse than useless – because damage to the reserve may be done and effort and resources wasted. Most modifications to management are likely to be changes in the times of treatment, substitution of one method for another and so on. The changes will be of degree, duration and timing rather than of actual type of management. This is where the importance of keeping detailed records is seen.

Simple Experimentation in Management

Almost all management for nature conservation consists of *trials,* because the concept of management for this purpose is relatively new. In almost no case can we say that such and such a regime of management is quite definitely indicated by the objectives of management. Thus experience of particular management regimes contributes to the sum of knowledge of management methods. In this section, however, we are concerned with more formal attempts to add to the knowledge on management which already exists.

There is a tendency to regard sophisticated *methods* as being of value in themselves. This is not only unsound thinking but creates unnecessary difficulties for those whose opportunities and resources are limited. Simple observations, provided they are well and objectively recorded, are still of importance, for even if the observer is unable to continue his observations his results may lead to further studies being made, as well as being of use in themselves. There is no substitute for practical experience in management for nature conservation as in many other fields, and if such experience is made available in an easily understood form, all conservationists can benefit.

In setting up a programme of observations, the observer would naturally wish to obtain maximum returns for the effort put in; and he would want any inferences made from the observations to be logically sound and efficient. He is well advised therefore to consult a statistician at an early preparatory stage of the project.

After simple observations the most useful way in which a contribution can be made to knowledge of management techniques is through a straightforward comparison of management methods. There is still much to be learnt on the effects that different treatments have on different animals and plants in varying situations and how one method compares with another when other factors, such as climate, weather, soil, aspect and slope, are similar. Enough is known already to show that animals and plants often have very different responses to management methods and that even within these two major groups different species may have quite different requirements for their conservation. Indeed, at the present time many objectives of management lie essentially in keeping the options open at all levels for the conservation of animal and plant species. Straightforward comparison between two methods or regimes of management applied to two areas of a reserve are open to a theoretical objection.

It must be borne in mind that in nature biological variation is so marked that it is not possible to assert with any great confidence that the observed differences are likely to have been caused by the different management. The observed difference may well be a chance phenomenon. Such observational programmes by conservationists are nevertheless valuable in suggesting interesting hypotheses and possibilities which can form the subject matter of a subsequent scientific investigation with proper replication.

In order to obtain statistically reliable results recourse must be made to the *field experiment*. For all except those with great resources of land, time, and staff the simple field experiment is to be preferred to more elaborate lay-outs. It is almost certainly true that more elaborate experiments can be made with plants than with animals. Quite small plots can be used satisfactorily with plants that could never be used when studying the effects of treatments on animals, or, of course, in investigating the effects of grazing animals on vegetation.

The study of the effects of certain *treatments* on vegetation or fauna is made in a *field trial* using the principles of *randomization* and *replication* to enable the estimation of the treatment effects and the *standard errors* of these estimates by estimating the magnitude of random variation, present in all biological situations. In the analysis of data from well designed field experiments, the total variation in the observations can be partitioned into variation due to treatments, due to blocks or other replications, and due to the inherent variability in the biological variable being measured (residual variation). Quite appropriately, this powerful technique is called the *Analysis of Variance*.

There are several kinds of *design* for field experiments. If there are '*a*' treatments and '*b*' replicates, the experiment must consist of $a \times b$ plots. If all plots are similar, the experimenter may use a *completely randomized design*. In this case each treatment will be allocated to b plots chosen at random (by tables of random numbers, drawing marked cards etc.). The number of treatments to be investigated and the number of replicates, or complete sets of treatments, are decided upon and plots of a convenient size marked out. There may be a theoretical minimum size for the plots, which will certainly have maximum and minimum sizes from a practical point of view. An 'untreated treatment' or *control* (check) is often necessary, but in a dynamic situation like grassland an unmanaged 'control' may be regarded as unnecessary if some form of treatment is essential for the maintenance of the grassland. On the other hand, an untreated control is usually necessary when comparing the results of fertilizers, for instance. A control, if present, counts as a treatment, although in some cases some form of untreated grassland may be obtained by enclosing small areas within the treated plots. Such areas are not, however, 'control plots' in the formal sense. A blank design of plots having been laid out, the different treatments are allocated to the plots at random.

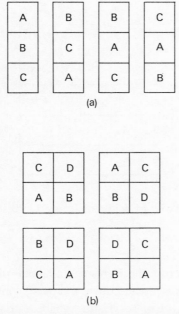

Fig. 12.1 Examples of layout of field plots using randomized blocks.
(a) four replicates of three treatments
(b) four replicates of four treatments

A simple and flexible design with restricted randomization is a *randomized block design*. If the experimental study area is not uniform (i.e. if all plots are not similar) the rationale is to partition the study area into *blocks* of similar plots, each treatment being applied once in each block. Thus, each block will have as many plots as there are treatments. Hence, each block is a complete replicate (Fig. 12.1).

More sophisticated, but less flexible, designs for field experiments exist. In the *Latin square* design the number of replicates must equal the number of treatments (Fig. 12.2). In this type of design, as can be seen from

A	B	C	D
B	A	D	C
C	D	A	B
D	C	B	A

(a)

A	B	C	D	E
C	E	D	B	A
E	D	A	C	B
B	C	E	A	D
D	A	B	E	C

(b)

Fig. 12.2 Examples of layout of field plots using Latin squares.
(a) 4 × 4
(b) 5 × 5

the figure, each *row* (across the page) contains one plot of each treatment and similarly each *column* (up and down the page) contains a plot with each treatment. For certain purposes Latin square and Graeco–Latin square designs are more efficient than randomized blocks, but a discussion of these is outside the scope of this book; details may be obtained from Fisher (1937) Finney (1960), Pearce (1953) and many other texts on statistical design.

As mentioned earlier, data from field experiments are often analysed by carrying out an analysis of variance. Although for a given experimental design the sequence of calculations for an analysis of variance can be made using a 'recipe book' approach this is, of course, dangerous in that the experimenter is not aware of what he is doing and gets no insight into the analysis of his experiment, but merely an 'answer' at the end of it. The likelihood of making mistakes, not just errors in calculation, through ignorance of statistical principles is also very great. Those conservationists without access to personal statistical advice, or those who wish to learn something of the principles of statistical method, can do much by reading and calculation of results. Bailey (1959), Campbell (1967) and Snedecor and Cochran (1967) can be recommended, although the last is expensive. In some cases the simplified approach and need for lucidity in reaching non-mathematical readers leads to over-simplification and a glossing-over of fundamental issues. Brownlee (1949) is very useful in giving detailed instructions for the analysis of many different kinds of design.

Conservationists who are interested in simple experimentation may find that they need to understand the principles of *factorial experiments* and how to deal with missing plots. Factorial experiments are ones which aim to analyse the effects of different *kinds* of treatment (or factors), such as mowing on the one hand and fertilizer application on the other, different *levels* of treatment (e.g. different rates of fertilizer application, different number of times of mowing), and interactions between factors. The standard texts deal with factorial experiments; a more advanced work is Yates (1937).

Inexperienced workers often assume that damage to an experiment necessitates its abandonment. In fact, techniques exist for the analysis of experiments where one plot, or even more, has been damaged, or untreated, or unrecorded. Such 'missing plot techniques' are explained in the textbooks whose titles have been quoted.

In experimental work it is often found, or suspected, that there is a relationship between one set of data (or variable) and another. For instance, a relationship between density of some kind of animal and the amount of vegetation present may be expected. Such possibilities are often examined by using *regression analysis*. Very often it is assumed that there is a straight line relationship between the two variables over the range observed, so that, if the two variables are x and y, $y = a + bx$. In *linear regression analysis* this assumption is examined and a and b are estimated.

For any fairly complicated statistical analyses, such as analysis of variance and regression analysis, it is almost essential to have access to a calculating machine. In any major computational job, an interested statistician is likely to be of considerable help.

Those conservationists who do experiments and obtain interesting and useful results should make them available to others interested in the same field of study. This should mean the publication of a scientific paper in a journal accessible to a wide range of interested people. Often, the local natural history society's Proceedings, or a County Trust's publication, will be the appropriate journal. More generally applicable information may find a place in more widely distributed journals. Papers with an agricultural flavour may be appropriate for the *Journal of The British Grassland Society*. Those more definitely conservationist in content could appear in *Biological Conservation*. More scientific and original papers may be accepted for *Oikos, Oecologia* or the *Journal of Applied Ecology*. Papers dealing primarily with one group of plants or animals may be most appropriately placed in specialist magazines or journals. Great care should be taken in the preparation of a paper for publication. The appropriateness of the paper for a particular journal should be determined and the style and 'instructions to authors' examined and studied in detail. All journals have their own idiosyncrasies and an adherence to the instructions given will save the author and editor much unnecessary work. Guidance on writing, preparing and illustrating manuscripts can be obtained from more experienced workers or from the many written guides which have been published.

References

ADAMSON, R. S. (1932), 'Notes on the natural regeneration of woodland in Essex', *J. Ecol.*, **20**, 152–156.

ALDHOUS, J. R. (1969, 'Chemical control of weeds in the forest', *Leafl. For. Commn*, no. 51.

ALDOUS, A. E. (1930), 'Effect of different clipping treatments on the yield and vigor of prairie grass vegetation, *Ecology*, **11**, 752–759.

ALLEE, W. C., EMERSON, A. E., PARK, O., PARK, T. and SCHMIDT, K. P. (1949), *Principles of animal ecology*, Saunders, Philadelphia and London.

ANDERSON, V. L. (1927), 'Studies on the vegetation of the English chalk. V. The water economy of the chalk flora', *J. Ecol.*, **15**, 72–129.

ANDRZEJEWSKA, L. (1962), '*Macrosteles laevis* Rib. as an unsettlement index of natural meadow association of Homoptera', *Bull. Acad. pol. Sci. Cl. II Ser. Sci. biol.*, **10**, 221–226.

ANDRZEJEWSKA, L. (1965), 'Stratification and its dynamics in meadow communities of Auchenorhyncha (Homoptera)', *Ekol. pol.*, (A) **13**, 685–715.

ANON. (1954), 'The calculation of irrigation need', *Tech. Bull. Minist. Agric. Fish. Fd*, no. 4.

ANSTED, D. T. (1865), *Applications of geology to the arts and manufactures*, Robert Hardwicke, London.

ARNOLD, G. W. (1960), 'Selective grazing by sheep of two forage species at different stages of growth', *Aust. J. agric. Res.*, **11**, 1026–1033.

ARNOLD, G. W. (1962), *The diet of grazing sheep*. Ph.D. thesis, London University.

ARNOLD, G. W. (1964), 'Factors within plant associations affecting the behaviour and performance of grazing animals', *Symp. Br. ecol. Soc.*, no. 4, 133–154.

ASH, J. S. (1970), 'Observations on a decreasing population of red-backed shrikes', *Br. Birds*, **63**, 185–225.

ASHMOLE, M. J. (1962), 'The migration of European thrushes; a comparative study based on ringing recoveries', *Ibis*, **104**, 314–346 and 522–559.

BAILEY, N. T. J. (1959), *Statistical methods in biology*, English Universities Press, London.

BAKER, H. (1937), 'Alluvial meadows: a comparative study of grazed and mown meadows', *J. Ecol.*, **25**, 408–420.

BAKER, S. (1973), *Milk to market*, Heinemann, London.

BARNES, H. F. (1946), *Gall midges of economic importance. 2. Gall midges of fodder crops*, Crosby Lockwood, London.

BATES, G. H. (1935), 'The vegetation of footpaths, sidewalks, cart-tracks and gateways', *J. Ecol.*, **23**, 470–487.

BEDDOWS, A. R. (1969), 'A history of the introduction of Timothy and Cocksfoot into alternate husbandry in Britain', *J. Br. Grassld Soc.*, **23**, 317–321; **24**, 40–44, 163–167.

BEIRNE, B. P. (1952), *British Pyralid and Plume Moths*, Warne, London.

BENSON, R. B. (1951, 1952, 1958), 'Symphyta'. *Handbk Ident. Br. Insects*, VI (2).

BERESFORD, M. W. and HURST, J. G. (eds) (1971), *Deserted medieval villages*, Lutterworth Press, London.

BEVAN, G. (1964), 'The feeding sites of birds in grassland with thick scrub. Some comparisons with dense oakwood', *Lond. Nat.*, **43**, 86–109.

BLACKWOOD, J. and TUBBS, C. R. (1970), 'A quantitative survey of chalk grassland in England', *Biol. Conserv.*, **3**, 1–5.

BOULET, L. J. (1939), *The ecology of a Welsh mountain sheep's walk*. Unpublished thesis, University College of Wales, Aberystywyth.

BOURNÉRIAS, M. (1968), '*Guide des groupements végétaux de la région Parisienne*', Société d'Edition d'Enseignement Supérieur, Paris.

BOWDEN, P. J. (1967), 'Agricultural prices, farm profits and rents', in *Agrarian history of England and Wales*, ed. Thirsk, J. **4**, 593–695.

BOYCOTT, A. E. (1934), 'The habitats of land Mollusca in Britain', *J. Ecol.*, **22**, 1–38.

BRADSHAW, A. D., LODGE, R. W., JOWETT, D. and CHADWICK, M. J. (1958), 'Experimental investigations into the mineral nutrition of several grass species. I. Calcium', *J. Ecol.*, **46**, 749–757.

BRENCHLEY, W. E. (1958), *The Park Grass Plots at Rothamsted 1856–1949*, Rothamsted Experimental Station, Harpenden, (Herts.)

BRIDGES, E. M. (1970), *World soils*, Cambridge University Press, Cambridge.

BRISTOWE, W. S. (1939), *The comity of spiders*, Vols I and II, Ray Society, London.

BROWNLEE, K. A. (1949), *Industrial experimentation*, H.M.S.O., London.

BUCKLAND, P. C. and KENWARD, H. K. (1973), 'Thorne Moor: a palaeo-ecological study of a Bronze Age site', *Nature, Lond.*, **241**, 405–406.

BUNCE, R. G. H. (1970), 'The flora of some relict oak woods in Wester Ross', *Trans. Proc. bot. Soc. Edinb.*, **40**, 565–575.

BUNTING, A. H. and ELSTON, J. (1966), 'Water relations of crops and grass on chalk soil', *Scient. Hort.*, 18, 116–120.

BURGES, A. and RAW, F. (1967), *Soil biology*, Academic Press, London.

CABORN, J. M. (1965), *Shelterbelts and windbreaks*, Faber and Faber, London.

CAIRD, J. (1852), *English agriculture in 1850–1*, Longman, London.

CAMPBELL, R. C. (1967), *Statistics for biologists*, Cambridge University Press, Cambridge.

CAMPBELL, R. S. (1960), 'Use of fire in grassland management', in *Working parts on pasture and fodder development in tropical America*, F.A.O.

CANCELADO, R. and YONKE, T. R. (1970), 'Effect of prairie burning on insect populations', *J. Kans. ent. Soc.*, **43**, 274–281.

CHADWICK, M. J. (1960), '*Nardus stricta* L.' *J. Ecol.*, **48**, 255–267.

CHAMBERS, J. D. and MINGAY, G. E. (1966), *The agricultural revolution*, Batsford, London.

CHIPPENDALE, H. G. and MERRICKS, R. W. (1956), 'Gang-mowing and pasture management', *J. Br. Grassld Soc.*, **11**, 1–10.

CHITTY, D., PIMENTEL, D. and KREBS, C. J. (1968), 'Food supply of over-wintered voles', *J. Anim. Ecol.*, **37**, 113–120.

CHURCH, B. M. and WEBBER, J. (1971), 'Fertiliser practice in England and Wales', *J. Sci. Fd Agric.*, **22**, 1–7.

CLARIDGE, M. F. (1959), 'The identity of *Eurytoma appendigaster* (Swederus, 1795) (Hymn., Eurytomidae), together with descriptions of some closely allied species bred from Gramineae', *Entomologist's mon. Mag.*, **95**, 2–13.

CLARIDGE, M. F. (1961), 'A contribution to the biology and taxonomy of some palaearctic species of *Tetramesa* (Walker (= *Isosoma* Walk = *Harmolita* Motsch.) (Hymenoptera: Eurytomidae), with particular reference to the British fauna', *Trans. R. ent. Soc. Lond.*, **113**, 175–216.

CLEMENTS, F. E. (1916), 'Plant succession: analysis of the development of vegetation', *Publs Carnegie Instn*, **242**, 1–512.

CLYMO, R. (1962), 'An experimental approach to part of the calcicole problem', *J. Ecol.*, **50**, 707–731.

CODY, M. L. (1968), 'On the methods of resource division in grassland bird communities', *Am. Nat.*, **102**, 107–147.

COLES, C. L. (1968), *Game conservation in a changing countryside: a guide for all shooters*, Museum Press, London.

COLQUHOUN, M. K. and MORLEY, AVERIL (1941), 'The density of downland birds', *J. Anim. Ecol.*, **10**, 35–46.

COMBER, THOMAS (1772), *Real improvements in agriculture*, Nicoll, London.

COOPE, G. R. (1968), 'Coleoptera from the "Arctic Bed" at Barnwell Station, Cambridge', *Geol. Mag.*, **105**, 482–486.

COOPE, G. R. (1969), 'The response of Coleoptera to gross thermal changes during the Mid Weichselian interstadial', *Mitt. int. Verein. theor. angew. Limnol.*, **17**, 173–183.

COOPE, G. R. (1970), 'Interpretation of quaternary insect fossils', *A. Rev. Ent.*, **15**, 97–120.

COOPE, G. R. and OSBORNE, P. J. (1967), 'Report on the coleopterous fauna of the Roman well at Barnsley Park, Gloucestershire', *Trans. Bristol Gloucs. archaeol. Soc.*, **86**, 84–87.

COPPOCK, J. T. (1962), Chapters 1–3 and 5, in *Changing use of land in Britain*, eds Best, R. H. and Coppock, J. T., Faber, London.

COPPOCK, J. T. (1964), *An agricultural atlas of England and Wales*, Faber, London.

CORNISH, M. W. (1954), 'The origin and structure of the grassland types of the central North Downs', *J. Ecol.*, **42**, 359–374.

CRAMPTON, C. B. and GARRETT JONES, R. (1967), 'The altitudinal zoning of grasslands and soils on the Brecon Beacons in South Wales', *J. Br. Grassld Soc.*, **22**, 46–52.

CRAWFORD, C. S. and HARWOOD, R. F. (1964), 'Bionomics and control of insects affecting Washington grass seed fields', *Tech. Bull. agric. Exp. Stn Wash. St.*, **44**, 25 pp.

CROMPTON, G. (1972), *History of Lakenheath Warren: an historical study for ecologists*. Unpublished report, The Nature Conservancy, London.

CRUTCHLEY, J. (1794), *General view of the agriculture of the county of Rutland*. Board of Agriculture, London.

CURTIS, J. T. and PARTCH, M. L. (1950), 'Some factors affecting flower production in *Andrapozon gerardi*', *Ecology*, **31**, 488–489.

DANKS, H. V. (1971), 'Populations and nesting-sites of some aculeate Hymenoptera nesting in *Rubus*', *J. Anim. Ecol.*, **40**, 63–77.

DARBY, H. C. (1952), 'The clearing of the English woodlands', *Geography*, **36**, 71–83.

DARBY, H. C. (1964), 'The draining of the English clay-lands', *Geogr. Z.*, **52**, 190–201.

DARBY, H. C. and CAMPBELL, E. M. J. (eds) (1962), *The Domesday geography of South-east England*. Cambridge University Press, Cambridge.

DAUBENMIRE, R. F. (1948), *Plants and environment*, John Wiley, New York.

DAVIES, H. T. (1969), 'Manuring of meadows in the Pennines', *Expl Husb.*, **18**, 8–24.

DAVIES, W. (1925), 'Relative palatability of pasture plants', *J. Minist. Agric. Fish.*, **32**, 106–116.

DAVIES, W. (1941), 'The grassland map of England and Wales: explanatory notes', *Agriculture, Lond.*, **48**, 112–121.

DAVIES, W. (1952), *The grass crop*, Spon, London.

DAVIS, THOMAS (1794), *General view of agriculture of the county of Wiltshire*, Board of Agriculture, London.

DEMPSTER, J. P. (1963), 'The population dynamics of grasshoppers and locusts', *Biol. Rev.*, **38**, 490–529.

DEMPSTER, J. P. (1971a), 'The population ecology of the Cinnabar Moth, *Tyria jacobaeae* L. (Lepidoptera, Arctiidae)', *Oecologia*, **7**, 26–67.

DEMPSTER, J. P. (1971b), 'Some effects of grazing on the population ecology of the Cinnabar Moth (*Tyria jacobaeae*)', *Symp. Br. ecol. Soc.*, no. 11, 517–526.

DEPARTMENT OF AGRICULTURE FOR SCOTLAND (1972), *Agricultural statistics, 1971, Scotland*. H.M.S.O., Edinburgh.

DE VOS, A. (1969), 'Ecological conditions affecting the production of wild herbivorous mammals on grassland', *Adv. ecol. Res.*, **6**, 137–183.

DIMBLEBY, G. (1967), *Plants and archaeology*, Baker, London.

DONISTHORPE, H. ST. J. K. (1927), *Guests of British ants*, Routledge, London.

DRUCE, G. C. (1886), *The flora of Oxfordshire*, Parker, London.

DUFFEY, E. (1962a), 'A population study of spiders in limestone grassland. Description of study area, sampling methods and population characteristics', *J. Anim. Ecol.*, **31**, 571–599.

DUFFEY, E. (1962b), 'A population study of spiders in limestone grassland. The field-layer fauna', *Oikos*, **13**, 15–34.

DUFFEY, E. (1965), 'The distribution of some rare Breckland spiders', *Trans. Suffolk Nat. Soc.*, **13** (2), 67–78.

DUFFEY, E. (1966), 'Spider ecology and habitat structure (Arach., Araneae)', *Senckenberg. biol.*, **47**, 1, 45–49.

DUFFEY, E. (1968), 'Ecological studies on the Large Copper Butterfly *Lycaena dispar*. Haw. *batavus* Obth. at Woodwalton Fen National Nature Reserve, Huntingdonshire', *J. appl. Ecol.*, **5**, 69–96.

DUFFEY, E. (in press), 'Lowland grassland and scrub: management for wildlife', in *Conservation in practice*, eds Goldsmith, B. and Warren, A., Wiley, London.

DUFFEY, E. and MORRIS, M. G. (1966), 'The invertebrate fauna of the chalk and its scientific interest', *Handbk a. Rep. Soc. Promot. Nat. Reserves*, 1966, 83–94.

DUNNET, G. M. and PATTERSON, I. J. (1968), 'The rook problem in North-east Scotland', in *The problems of birds as pests*, eds Murton, R. K. and Wright, E. M., Academic Press, London.

EADY, R. D. and QUINLAN, J. (1963), 'Hymenoptera Cynipoidea', *Handbk Ident. Br. Insects*, **8** (1a), 81 pp.

EASTON, A. M. (1967), 'The Coleoptera of a dead fox (*Vulpes vulpes* (L.)); including two species new to Britain', *Entomologist's mon. Mag.*, **102**, (1966), 205–210.

EASTOP, V. F. (1972), 'Deductions from the present day host plants of aphids and related insects', *Symp. R. ent. Soc. Lond.*, no. 6, 157–178.

EGGELING, W. J. (1964), 'A nature reserve management plan for the Island of Rhum, Inner Hebrides', *J. appl. Ecol.*, **1**, 405–419.

EICHHORN, O. (1967), 'Insects attacking rose hips in Europe', *Tech. Bull. Commonw. Inst. biol. Control*, **8**, 83–102.

ELLENBERG, H. (1963), *Vegetation Mitteleuropas mit den Alpen*. Eugen Ulmer, Stuttgart.

ELLENBERG, H. and MUELLER-DOMBOIS, D. (1966a), Tentative physiognomic-ecological classification of plant formations of the earth', *Ber. geobot. ForschInst. Rübel*, **37**, 21–55.

ELLENBERG, H. and MUELLER-DOMBOIS, D. (1966b), 'A key to Raunkiaer plant life forms with revised subdivisions', *Ber. geobot. ForschInst. Rubel*, **37**, 56–73.

ELLIOTT, R. J. (1953), *Heather burning*, Ph.D. thesis, Sheffield University.

ELLISON, L. (1946), 'The pocket gopher in relation to soil erosion on mountain range', *Ecology*, **27**, 101–114.

ELTON, C. S. (1927), *Animal ecology*, Sidgwick and Jackson, London.

ELTON, C. S. (1966), *The pattern of animal communities*, Methuen, London.

ELTON, C. S. and MILLER, R. S. (1954), 'The ecological survey of animal communities, with a practical system of classifying habitats by structural characters', *J. Ecol.*, **42**, 160–196.

EVANS, J. G. (1972), *Land snails in archaeology*, Seminar Press, London and New York.

EVERSON, A. C. (1966), 'Effects of frequent clipping at different stubble heights on Western Wheatgrass (*Agropyron smithii* Rydb.)', *Agron. J.*, **58**, 33–35.

EYRE, S. R. (1955), 'The curving plough-strip and its historical implications', *Agric. Hist. Rev.*, **3**, 80–94.

FARRER, F. E. (1936), 'A moveable fee simple lot meadow', *Conveyancer*, **1** (New Series), 53–61.

FARROW, E. P. (1915), 'On the ecology of the vegetation of Breckland', *J. Ecol.*, **3**, 211–228.

FENTON, E. W. (1937), 'The influence of sheep on the hill grazings in Scotland', *J. Ecol.*, **25**, 424–430.

FESTETICS, A. and LEISLER, B. (1968), 'Ecology of waterfowl in the region of Lake Neusiedl, Austria, particularly in the World Wildlife Fund Seewinkel Reserve', *Wildfowl*, **19**, 83–95.

FEWKES, D. W. (1961), 'Diel vertical movements in some grassland Habidae (Heteroptera)', *Entomologist's mon. Mag.*, **97**, 128–130.

FINNEY, D. J. (1960), *Introduction to theory of experimental design.* University of Chicago Press, Chicago.

FISHER, J. P. (1970), 'The biology and taxonomy of some chalcidoid parasites (Hymenoptera) of stem-living larvae of *Apion* (Coleoptera: Curculionideae)', *Trans. R. ent. Soc. Lond.*, **122**, 293–322.

FISHER, R. A. (1937), *The design of experiments*, Oliver and Boyd, Edinburgh.

FISHER, R. A. and YATES, F. (1959), *Statistical tables for biological, agricultural and medical research*, Oliver and Boyd, Edinburgh.

FITTER, A. (1968), 'The present distribution of juniper (*Juniperus communis*) in the Chilterns', *Proc. Rep. Ashmol. nat. Hist. Soc. Oxf.*, 1968, 16–23.

FITZHERBERT, ANTHONY (1523), *Surveyinge.*

FLETCHER, T. W. (1961), 'Lancashire livestock farming in the Great Depression', *Agric. Hist. Rev.*, **9**, 17–42.

FORBES, E. (1846), 'On the connection between the distribution of the existing fauna and flora of the British Isles, and the geological changes which have affected their area, especially during the epox of the Northern Drift', *Mem. geol. Surv. U.K.*, **1**, 336–432.

FORD, L. T. (1949), *A guide to the smaller British Lepidoptera.* South London Entomological and Natural History Society, London.

FOREST SERVICE, U.S. DEPARTMENT OF AGRICULTURE (1948), *Woody-plant seed manual*, U.S. Department of Agriculture, Washington.

FRAZER, J. F. D. (1961), 'Butterfly populations on the North Downs', *Proc. Trans. S. Lond. ent. nat. Hist. Soc.*, 1960, 89–109.

FREEMAN, B. E. (1965), 'An investigation of the distribution during winter of the white clover seed weevil, *Apion dichroum* Bedel (Col., Curculionidae)', *J. appl. Ecol.*, **2**, 105–113.

FRYER, D. J. and MAKEPEACE, R. (eds) (1972), *British Crop Protection Council Weed Control Handbook Vol. II Recommendations*, Seventh edition, Blackwell Scientific Publications, Oxford.

FRYER, D. W. (1941), 'Huntingdonshire', in *The land of Britain*, ed. Stamp, L. D. **75**, 411–454, Geographical Publications, London.

FUSSELL, G. E. (1964), 'The grasses and grassland cultivation of Britain, II. 1700–1900'. *J. Br. Grassld Soc.*, **19**, 49–54, 212–217.

GARREN, K. H. (1943), 'Effects of fire on the vegetation of the south-eastern United States', *Bot. Rev.*, **9**, 617–654.

GAY, P. A., GREEN, B. H. and LABERN, M. V. (1968), 'The experimental management of chalk heath at Lullington Heath National Nature Reserve, Sussex', *J. Ecol.*, **56**, 240–250.

GEIGER, R. (1965), *The climate near the ground*, Harvard University Press, Cambridge, Mass.

GIMINGHAM, C. H. (1971), '*Calluna* heathlands: use and conservation in the light of some ecological effects of management', *Symp. Br. ecol. Soc.*, no. 11, 91–103.

GIMINGHAM, C. H. (1972), *Ecology of heathlands*, Chapman and Hall, London.

GODFREY, G. K. (1953), 'The food of *Microtus agrestis hirtus* (Bellamy) in Wytham, Berkshire', *Säugetierk. Mitt.*, **1**, 148–151.

GODFREY, G. K. (1955), 'Observations on the nature of the decline in numbers of *Microtus* populations', *J. Mammal.*, **36**, 209–241.

GODWIN, H. (1936), 'Studies in the ecology of Wicken Fen. III. The establishment and development of fen scrub (carr)', *J. Ecol.*, **24**, 82–116.

GODWIN, H. (1944), 'The age and origin of the "Breckland" heaths of East Anglia', *Nature, Lond.*, **154**, 6.

GODWIN, H. (1956), *The history of the British flora*, Cambridge University Press, Cambridge.

GODWIN, H. (1967), 'The ancient cultivation of hemp', *Antiquity*, **41**, 42–50.

GODWIN, H. and WILLIS, E. H. (1962), Cambridge University natural radiocarbon measurements, V', *Radiocarbon*, **4**, 57–70.

GOOGE, BARNABY (1577), *Foure Bookes of husbandry*.

GOULD, S. J. and JOHNSON, R. F. (1972), Geographic variation. *Annu. Rev. Ecol. & Syst.*, **3**, 457–498.

GRANT, S. A. and HUNTER, R. F. (1968), 'Variation in yield, maturity type, winter greenness and sensitivity to cutting of hill grass species', *J. Br. Grassld Soc.*, **23**, 149–155.

GREEN, B. H. (1972), 'The relevance of seral eutrophication and plant competition to the management of successional communities', *Biol. Conserv.*, **4**, 378–384.

GREENSLADE, P. J. M. (1964), 'Pitfall trapping as a method for studying populations of Carabidae (Coleoptera)', *J. Anim. Ecol.*, **33**, 301–310.

GREIG-SMITH, P. (1964), *Quantitative plant ecology*, Butterworth Scientific Publications, London.

GRETTON, R. H. (1912), 'Historical notes on the lot-meadow customs at Yarnton, Oxon', *Econ. J.*, **22**, 53–62.

GRIME, J. P., LOACH, K. and PECKHAM, D. (1971), 'Control of vegetation succession by means of soil fabrics', *J. appl. Ecol.*, **8**, 257–263.

GRINELL, J. (1924), 'Geography and evolution', *Ecology*, **5**, 225–229.

GRUBB, P. H., GREEN, H. E. and MERRIFIELD, R. C. J. (1969), 'The ecology of chalk heaths: its relevance to the calcicole–calcifuge and soil acidification problems', *J. Ecol.*, **57**, 175–212.

HALE, W. G. (1967), 'Collembola', in *Soil biology*, eds Burges, A. and Raw, F., 397–411. Academic Press, London.

HANSSON, L. (1968), 'Population densities of small mammals in open-field habitats in south Sweden in 1964–1967', *Oikos*, **19**, 53–60.

HANSSON, L. (1971), 'Small rodent food, feeding and population dynamics. A comparison between granivorous and herbivorous species in Scandinavia', *Oikos*, **22**, 183–198.

HARLEY, J. B. (1964), *The historian's guide to Ordnance Survey maps*, National Council for Social Services for Standing Conference for Local History, London.

HARPER, J. L. and WHITE, J. (1971), 'The dynamics of plant populations', *Proc. Adv. Study Inst. Dynamics Numbers Popul.*, Oosterbeek, 1970, 41–63.

HARRIS, A. (1961), *The rural landscape of the East Riding*, Oxford University Press, Oxford.

HARTLEY, P. H. T. (1954), 'Wild fruits in the diet of British thrushes. A study in the ecology of closely allied species', *Br. Birds*, **47**, 97–107.

HARVEY, P. D. A. and SKELTON, R. A. (1969), 'Medieval English maps and plans', *Imago Mundi*, **23**, 101–102.

HASLAM, S. M. (1965), 'Ecological studies in the Breck Fens. I. Vegetation in relation to habitat', *J. Ecol.*, **53**, 599–619.

HEATH, J. (ed.) (1970), *Provisional atlas of the insects of the British Isles*. I. *Lepidoptera Rhopalocera butterflies*. The Nature Conservancy, Monks Wood Experimental Station.

HEDDLE, R. G. (1967), 'Long-term effects of fertilizers on herbage production. I. Yields and botanical composition', *J. agric. Sci., Camb.*, **69**, 425–431.

HERING, M. (1951), *Biology of the leaf miners*, Junk, 's-Gravenhage.

HERRIOTT, J. B. D. and WELLS, D. A. (1963), 'The grazing animal and sward productivity', *J. agric. Sci., Camb.*, **61**, 89–99.

HOOPER, M. D. (1968), 'The conservation of plants', in *Hedges and hedgerow trees*, eds Hooper, M. D. and Holdgate, M. W., Monks Wood Experimental Station Symposium No. 4, 50–52. Nature Conservancy, London.

HOOPER, M. D. (1970), 'Dating hedges', *Area*, **4**, 63–65.

HOOPER, M. D. and HOLDGATE, M. W. (1968), 'Developments since the symposium. 2. The costs of hedgerows and other fences', in *Hedges and hedgerow trees*, Monks Wood Experimental Station Symposium No. 4, 95–96. Nature Conservancy, London.

HOPE-JONES, P. (1972), 'Succession in breeding bird populations of sample Welsh oakwoods', *Br. Birds*, **65**, 291–299.

HOPE-SIMPSON, J. F. (1940), 'Studies of the vegetation of the English chalk. VI. Late stages in succession leading to chalk grassland', *J. Ecol.*, **28**, 386–397.

HOPE-SIMPSON, J. F. (1941), 'Studies of the vegetation of the English chalk. VIII. A second survey of the chalk grasslands of the South Downs', *J. Ecol.*, **29**, 217–267.

HOSKINS, W. G. (1957), *Midland peasant*, Macmillan, London.

HOSKINS, W. G. (1964), 'Harvest fluctuations and English economic history, 1480–1619', *Agric. Hist. Rev.*, **12**, 28–46.

HOSKINS, W. G. (1968), 'Harvest fluctuations and English economic history, 1620–1759', *Agric. Hist. Rev.*, **16**, 15–31.

HUGHES, R. ELFYN, DALE, J., ELLIS WILLIAMS, I. and REES, D. I. (1973), 'Studies in sheep population and environment in the mountains of north-west Wales', *J. appl. Ecol.*, **10**, 113–132.

HUGHES, R. J. (1955), 'The influence of the prevailing weather on the numbers of *Meromyza variegata* Meigen (Diptera, Chloropidae), caught with a sweep net', *J. Anim. Ecol.*, **24**, 324–335.

HUNTER, R. F. (1962), 'Hill sheep and their pasture: a study of sheep grazing in south-east Scotland', *J. Ecol.*, **50**, 651–680.

HUNTER, R. F. (1964), 'Home range behaviour in hill sheep', *Symp. Br. ecol. Soc.*, no. 4, 155–171.

HUTCHINSON, G. E. (1945), 'Aluminium in soils, plants and animals', *Soil Sci.*, **60**, 29–40.

HUTCHINSON, G. E. (1957), 'Concluding remarks', *Cold Spring Harb. Symp. quant. Biol.*, **22**, 415–427.

IVENS, G. W. (1972), 'The problem of bush in rangeland', *Span*, **15**, 1, 23–26.

IVERSEN, J. (1949), 'The influence of prehistoric man on vegetation', *Danm. geol. Unders.*, **3**, 1–25.

IVIMEY-COOK, R. B. and PROCTOR, M. C. F. (1966), 'The plant communities of the Burren, Co. Clare', *Proc. R. Ir. Acad.*, B, **64**, 211–301.

IVINS, J. D. (1952), 'The relative palatability of herbage plants', *J. Br. Grassld Soc.*, **7**, 43–54.

JACKSON, G. and SHELDON, J. (1949), 'The vegetation of magnesian limestone cliffs at Marklands Grips near Sheffield', *J. Ecol.*, **37**, 38–50.

JANZEN, D. H. (1970), 'Herbivores and the number of tree species in tropical forests', *Am. Nat.*, **104**, 501–528.

JANZEN, D. H. (1971), 'Seed predation by animals', *Annu. Rev. Ecol. & Syst.*, **2**, 465–492.

JEFFERIES, R. L. and WILLIS, A. J. (1964), 'Studies on the calcicole–calcifuge habit', *J. Ecol.*, **52**, 691–707.

JEFFERS, J. N. R. (1972), 'The challenge of modern mathematics to the ecologist', *Symp. Br. ecol. Soc.*, no. 12, 1–11.

JONES, M. G. (1933), 'Grassland management and its influence on the sward', *Emp. J. exp. Agric.*, **1**, 43–57, 122–128, 223–234, 360–366, 366–367.

JOHNSON, J. and MEADOWCROFT, S. C. (1968), 'Manuring of permanent meadows 1953–61', *Expl Husb.*, **17**, 70–79.

KEEP, E. (1972), 'Variability in the wild raspberry', *New Phytol.*, **71**, 915–924.

KELLY, M. and OSBORNE, P. J. (1964), 'Two faunas and floras from the alluvium at Shustoke, Warwickshire', *Proc. Linn. Soc. Lond.*, **176**, 37–65.

KERNEY, M. P., BROWN, E. H. and CHANDLER, T. J. (1964), 'The late-glacial and post-glacial history of the chalk escarpment near Brook, Kent', *Phil. Trans. R. Soc.*, B, **248**, 135–204.

KERRIDGE, E. (1967), *The agricultural revolution*, Allen and Unwin, London.

KEVAN, D. K. McE. (1962), *Soil animals*, Witherby, London.

KING, J. and NICHOLSON, I. A. (1964), 'Grasslands of the forest and sub-alpine zones', in *The vegetation of Scotland*, ed. Burnett, J. H., 168–231. Oliver and Boyd, Edinburgh.

KULLENBERG, B. (1946), 'Studien über die Biologie der Capsiden', *Zool. Bidr. Upps.*, **23**, 1–522.

KYDD, D. D. (1964), 'The effect of different systems of cattle grazing on the botanical composition of permanent downland pasture', *J. Ecol.*, **52** 139–149.

LACK, D. (1935), 'The breeding bird population of British heaths and moorland', *J. Anim. Ecol.*, **4**, 43–51.

LANGER, R. H. M. (1956), 'Growth and nutrition of timothy. I. The life-history of individual tillers', *Ann. appl. Biol.*, **44**, 166–187.

LANGER, R. H. M. (1959), 'A study of growth in swards of timothy and meadow fescue. 2. The effects of cutting treatments', *J. agric. Sci., Camb.*, **52**, 273–281.

LARSSON, T. (1969), 'Land use and bird fauna on shores in southern Sweden', *Oikos*, **20**, 136–155.

LaVELLE, J. W., SEILHEIMER, J. A., OSBORN, N. L. and HERRMAN, S. J. (1969), 'A preliminary study of three lentic communities on the Pawnee national grasslands', in *The grassland ecosystem: a preliminary synthesis*, eds Dix, R. L. and Beidleman, R. G., Colorado State University, Colorado.

LAYCOCK, W. A. and CONRAD, P. W. (1969), 'How time and intensity of clipping affect Tall Bluebell', *J. Range Mgmt*, **22**, 299–303.

LE QUESNE, W. J. (1969), 'Hemiptera Cicadomorpha Deltocephalinae', *Handbk Ident. Br. Insects*, **2** (2b), 65–148.

LEWIS, T. and DIBLEY, G. C. (1970), 'Air movement near windbreaks and a hypothesis of the mechanism of the accumulation of airborne insects', *Ann. appl. Biol.*, **66**, 477–484.

LINDEMAN, R. L. (1942), 'The tropic-dynamic aspect of ecology', *Ecology*, **23**, 399–418.

LIPSCOMB, C. G. and JACKSON, R. A. (1964), 'Some considerations on some present-day conditions as they affect the continued existence of certain butterflies', *Entomologist's Rec. J. Var.*, **76**, 63–68.

LITTLE, E. (1845), 'The farming of Wiltshire', *Jl R. agric. Soc.*, 1st Series, **5**, 161–179.

LIVINGSTONE, R. B. (1972), 'Influence of birds, stones and soil on the establishment of pasture juniper, *Juniperus communis* and red cedar, *J. virgineana* in New England pastures', *Ecology*, **53**, 1141–1147.

LLOYD, P. S. (1968), 'The ecological significance of fire in limestone grassland communities of the Derbyshire dales', *J. Ecol.*, **56**, 811–826.

LLOYD, P. S. (1972), 'The grassland vegetation of the Sheffield Region. II. Classification of grassland types', *J. Ecol.*, **60**, 739–776.

LLOYD, P. S. and PIGOTT, C. D. (1967), 'The influence of soil conditions on the course of succession on the chalk of southern England', *J. Ecol.*, **55**, 137–146.

LUFF, M. L. (1965), 'The morphology and microclimate of *Dactylis glomerata* tussocks', *J. Ecol.*, **53**, 771–787.

LUFF, M. L. (1966), 'The abundance and diversity of the beetle fauna of grass tussocks', *J. Anim. Ecol.*, **35**, 189–208.

MACAN, T. T. (1963), *Freshwater ecology*, Longman, London.

MACAN, T. T. and MacFADYEN, A. (1941), 'The water bugs of dewponds', *J. Anim. Ecol.*, **10**, 175–183,

McNAUGHTON, S. J. and WOLF, L. L. (1970), 'Dominance and the niche in ecological systems', *Science, N.Y.*, **167**, 131–139.

McNEIL, S. (1971), 'The energetics of a population of *Leptopterna dolobrata* (Hereroptera: Miridae)', *J. Anim. Ecol.*, **40**, 127–140.

McVEAN, D. N. (1956), 'Ecology of *Alnus glutinosa* (L.) Gaertn. V. Notes on some British alder populations', *J. Ecol.*, **44**, 321–330.

McVEAN, D. N. and LOCKIE, J. D. (1969), *Ecology and land use in upland Scotland*, Edinburgh University Press, Edinburgh.

McVEAN, D. N. and RATCLIFFE, D. A. (1962), *Plant communities of the Scottish highlands*, H.M.S.O., London.

MALICKY, H., SOBHIAN, R. and ZWÖLFER, H. (1970), 'Studies in the possibilities for biological control of *Rhamnus cathartica* L. in Canada: host ranges, feeding sites and phenology of insects associated with European Rhamnacae', *Z. angew. Ent.*, **65**, 77–97.

MALLOCH, A. J. C. (1971), 'Vegetation of the maritime cliff tops of the Lizard and Lands End peninsulas, West Cornwall', *New Phytol.*, **70**, 1155–1197.

MARGALEF, R. (1963), 'On certain unifying principles in ecology', *Am. Nat.*, **97**, 357–374.

MARGALEF, R. (1968), *Perspectives in ecology theory*, University Press, Chicago.

MARTIN, D. J. (1964), 'Analysis of sheep diet utilising plant epidermal fragments in faeces samples', *Symp. Br. ecol. Soc.*, no. 4. 173–188.

MASSEE, A. M. (1954), *The pests of fruit and hops*, Crosby Lockwood, London.

MATTHEWS, J. R. (1937), 'Geographical relationships of the British flora', *J. Ecol.*, **25**, 1–90.

MAY, L. H. (1960), 'The utilisation of carbohydrate reserves in pasture plants after defoliation', *Herb. Abstr.*, **30**, 239–343.

MELLANBY, K. (1968), 'The effects of some mammals and birds on the regeneration of oak', *J. appl. Ecol.*, **5**, 359–366.

MELLANBY, K. (1971), *The mole*, Collins, London.

MERTON, L. F. H. (1970), 'The history and status of the woodlands of the Derbyshire limestone', *J. Ecol.*, **58**, 723–744.

MILLER, G. R. (1971), 'Grazing and the regeneration of shrubs and trees', in *Range ecology research*, ed. Nicholson, I. A., 27–40. The Nature Conservancy, Edinburgh.

MILLER, R. B. STOUT, J. B. and LEE, K. E. (1955), 'Biological and chemical changes following scrub burning on a New Zealand hill soil', *N.Z. Jl Sci. Technol.*, **37**, 290–313.

MILNER, C. and HUGHES, R. E. (1968), *Methods for the measurement of the primary production of grassland*, Blackwell Scientific Publications for the International Biological Programme, I.B.P. Handbook No. 6 Oxford.

MILTON, W. E. J. (1940), 'The effect of manuring, grazing and cutting on the yield, botanical and chemical composition of natural hill pastures. I. Yield and botanical section', *J. Ecol.*, **28**, 326–356.

MILTON, W. E. J. (1947), 'The yield, botanical and chemical composition of natural hill herbage under manuring, controlled grazing and hay conditions. I. Yield and botanical section', *J. Ecol.*, **35**, 65–89.

MINISTRY OF AGRICULTURE, FISHERIES AND FOOD (1968), *A century of agricultural statistics*, H.M.S.O., London.

MINISTRY OF AGRICULTURE, FISHERIES AND FOOD (1972), *Agricultural statistics, 1970/71*, H.M.S.O., London.

MOORE, I. (1966), *Grass and grasslands*, Collins, London.

MOORE, N. W. (1968), 'The conservation of animals', in *Hedges and hedgerow trees*, eds Hooper, M. D. and Holdgate, M. W., Monks Wood Experimental Station Symposium No. 4, 53–57. The Nature Conservancy, London.

MORETON, B. D. (1969), 'Beneficial insects and mites', *Bull. Minist. Agric. Fish. Fd, Lond.*, No. 20.

MORRIS, M. G. (1967a), 'Differences between the invertebrate faunas of grazed and ungrazed chalk grassland. I. Responses of some phytophagous insects to cessation of grazing', *J. appl. Ecol.*, **4**, 459–474.

MORRIS, M. G. (1967b), 'The representation of butterflies (Lep., Rhopalocera) on British statutory nature reserves', *Entomologist's Gaz.*, **18**, 57–68.

MORRIS, M. G. (1968), 'Differences between the invertebrate faunas of grazed and ungrazed chalk grassland. II. The fauna of sample turves', *J. appl. Ecol.*, **5**, 601–611.

MORRIS, M. G. (1969a), 'Populations of invertebrate animals and the management of chalk grassland in Britain', *Biol. Conserv.*, **1**, 225–231.

MORRIS, M. G. (1969b), 'Differences between the invertebrate faunas of grazed and ungrazed chalk grassland. III. The heteropterous fauna', *J. appl. Ecol.*, **6**, 475–487.

MORRIS, M. G. (1971a), 'Differences between the invertebrate faunas of grazed and ungrazed chalk grassland. IV. Abundance and diversity of Homoptera–Auchenorhyncha', *J. appl. Ecol*, **8**, 37–52.

MORRIS, M. G. (1971b), 'The management of grassland for the conservation of invertebrate animals', *Symp. Br. ecol. Soc.*, no. 11, 527–552.

MORRIS, M. G. (1973), 'The effects of seasonal grazing on the Heteroptera and Auchenorhyncha (Hemiptera) of chalk grassland', *J. appl. Ecol.*, **10**, 761–789.

MORRIS, M. G. (1974), 'Auchenorhyncha (Hemiptera) of the Burren, with special reference to species-associations of the grasslands', *Proc. R. Ir. Acad. B*, **74** 7–30.

MORRIS, R. M. and THOMAS, J. G. (1972), 'The seasonal pattern of dry-matter production of grasses in the North Pennines', *J. Br. Grassld Soc.*, **27**, 163–172.

MOSS, C. E. (1913), *Vegetation of the Peak District*, Cambridge University Press, Cambridge.

MURPHY, P. W. (ed.) (1962), *Progress in soil zoology*, Butterworth, London.

MURRAY, K. A. H. (1955), *Agriculture*, Longman, London.

MURTON, R. K. (1971), *Man and birds*, Collins, London.

NEW, T. R. (1971), 'An introduction to the natural history of the British Psocoptera', *Entomologist*, **104**, 59–97.

NICHOLSON, I. A. (1971), 'Some effects of animal grazing and browsing on vegetation', *Trans. Proc. bot. Soc. Edinb.*, **41**, 85–94.

NICHOLSON, I. A. PATERSON, I. S. and CURRIE, A. (1970), 'A study of vegetational dynamics: selection by sheep and cattle in *Nardus* pasture', *Symp. Br. ecol. Soc.*, no. 10, 129–143.

NIELSEN, C. OVERGAARD (1967), 'Nematoda', in *Soil biology*, eds Burges, A. and Raw, F., 197–211, Academic Press, London.

NORMAN, M. J. T. (1957), 'The influence of various grazing treatments upon the botanical composition of a downland permanent pasture', *J. Br. Grassland Soc.*, **12**, 246–256.

NORMAN, M. J. T. (1960), 'The relationship between competition and defoliation in pasture', *J. Br. Grassld Soc.*, **15**, 145–149.

NORMAN, M. J. T. and GREEN, J. O. (1958), 'The local influence of cattle dung and urine upon the yield and botanical composition of permanent pasture', *J. Br. Grassld Soc.*, **13**, 39–45.

OBERDORFER, E. (1970), *Pflanzensoziologische Exkursionsflora für Süddeutschland und die angrenzenden Gebiete*, Ulmer, Stuttgart.

ODUM, E. P. (1959), *Fundamentals of ecology*, Saunders, Philadelphia and London.

OLIVER, C. G. (1972), 'Genetic and phenotypic differentiation and geographic distance in four species of lepidoptera', *Evolution, Lancaster, Pa.*, **26**, 221–241.

ORDNANCE SURVEY (1952), 'Vegetation: the grasslands of England and Wales', *Explanatory Texts*, No. 5.

ORWIN, C. S. and WHETHAM, E. H. (1971), *History of the British agriculture 1846–1941*, David and Charles, Newton Abbot.

OSBORNE, P. J. (1969), 'An insect fauna of late Bronze Age date from Wilsford, Wiltshire', *J. Anim. Ecol.*, **38**, 555–566.

O'SULLIVAN, A. M. (1965), *A phytosociological survey of Irish lowland pastures*. Ph.D. thesis, University College, Dublin.

OWENSBY, C. E. and ANDERSON, K. L. (1969), 'Effect of clipping date on loamy Upland Bluesten range', *J. Range Mgmt*, **22**, 351–354.

PARNELL, J. R. (1962), *The insects living in the pods of broom* (Sarothamnus scoparius) *and the relationships between them*. Ph.D. thesis, University of London.

PASSARGE, H. (1969), 'Zur soziologischen Gliederung mitteleuropäischer Frischwiesen', *Reprium nov. Spec. Regni veg.*, **80**, 357–372.

PAWSON, H. C. (1960), *Cockle Park farm*, University Press, Oxford.

PEARCE, S. C. (1953), *Field experiment with fruit trees and other perennial plants*, Commonwealth Agricultural Bureaux, Farnham Royal.

PELL, A. (1899), *The making of the land in England: a retrospect*, Murray, London.

PENMAN, H. L. (1948), 'Natural evaporation from open water, bare soil and grass', *Proc. R. Soc.* A, **193**, 120–145.

PENMAN, H. L. (1952), 'Water and plant growth', *Agric. Prog.*, **27**, 147–154.

PENNINGTON, W. (1969), *The history of British vegetation*, English Universities Press, London.

PEPPER, H. W. and TEE, L. A. (1972), 'Forest fencing', *Forest Rec., Lond.*, **80**, 1–50.

PERRING, F. H. (ed.) (1968), *Critical supplement to the atlas of the British flora*, Botanical Society of the British Isles, London.

PERRING, F. H. and WALTERS, S. M. (1962), *Atlas of the British flora*, Thomas Nelson and Sons for the Botanical Society of the British Isles, London.

PERRY, P. J. (ed.) (1973), *British agriculture, 1875–1914*, Methuen, London.

PETERKEN, G. F. and HUBBARD, J. C. E. (1972), 'The shingle vegetation of southern England; the holly wood on Holmstone Beach, Dungeness', *J. Ecol.,* **60**, 547–572.

PHILLIPS, A. D M. (1972), 'The development of underdraining on a Yorkshire estate during the nineteenth century', *Yorks. archaeol. J.,* **44**, 195–206.

PIELOU, E. C. (1972), 'Niche width and niche overlap: a method for measuring them', *Ecology,* **53**, 687–692.

PIGOTT, C. D. (1970), 'The response of plants to climate and climatic change', in *The flora of a changing Britain*, ed. Perring, F. H., 32–44. E. W. Classey, Hampton, Middlesex.

PIGOTT, C. D. and PIGOTT, M. E. (1963), 'Late-glacial and post-glacial deposits at Malham, Yorkshire', *New Phytol.,* **62**, 317–334.

PIGOTT, C. D. and WALTERS, S. M. (1953), 'Is the box-tree a native of England?' in *The changing flora of Britain*, ed. Lousley, J. E., 184–187. Report of the 1952 Botanical Society of the British Isles Conference, Arbroath.

PIGOTT, C. D. and WALTERS, S. M. (1954), 'On the interpretation of the discontinuous distribution shown by certain British species of open habitats', *J. Ecol.,* **42**, 95–116.

PRINCE, H. C. (1959), 'The tithe surveys of the mid-nineteenth century', *Agric. Hist. Rev.,* **7**, 14–26.

PROCTOR, M. C. F. (1956), '*Helianthemum* Mill', *J. Ecol.,* **44**, 675–692 (Biological Flora of the British Isles).

RABOTNOV, T. A. (1950), 'Life cycles of perennial herbage plants in meadow communities', *Proc. Komarov Bot. Inst. Akad. Sci. U.S.S.R.,* Ser. 3(6), 7–240 (in Russian).

RABOTNOV, T. A. (1960), 'Some problems in increasing the proportion of leguminous species in permanent meadows', *Int. Grassld Congr.,* 8th, 260–264.

RABOTNOV, T. A. (1969), 'Plant regeneration from seed in meadows of the U.S.S.R.', *Herb. Abstr.,* **39**, 269–277.

RACKHAM, O. (1971), 'Historical studies and woodland conservation', *Symp. Br. ecol. Soc.,* no. 11, 563–580.

RANWELL, D. S. (1972a), *The management of sea buckthorn* Hippophaë *rhamnoides L. on selected sites in Great Britain*. Report of the Hiphophaë Study Group, The Nature Conservancy.

RANWELL, D. S. (1972b), *Ecology of salt marshes and sand dunes*, Chapman and Hall, London.

RAWES, M. and WELCH, D. (1969), 'Upland productivity of vegetation and sheep at Moor House National Nature Reserve, Westmorland, England', *Oikos,* Suppl. no. 11, 7–72.

RAUNKIAER, C. (1934), *The life forms of plants and statistical plant geography*, Clarendon Press, Oxford.

REDDINGIUS, J. (1971), 'Models as research tools', in *Dynamics of populations*, eds den Boer, P. J. and Gradwell, G. R., 64–76, Centre for Agricultural Publishing and Documentation, Wageningen.

REDFERN, M. (1968), 'The natural history of spear thistle heads', *Fld Stud.*, **2** (5), 669–717.

REID, C. (1911), 'Relation of the present plant population of the British Isles to the glacial period', *Rep. Br. Ass. Advmt Sci.*, Section K, 573–580.

RICHARDS, O. W. (1926), 'Animal communities of the felling and burn successions at Oxshott Heath, Surrey', *J. Ecol.*, **14**, 244–281.

RICHARDS, O. W. and WALOFF, N. (1954), 'Studies on the biology and population dynamics of British grasshoppers', *Anti-Locust Bull.*, **17**, 182 pp.

RIDLEY, H. M. (1930), *The dispersal of plants throughout the world*, 1–744. L. Reeve & Co. Ltd., Ashford, Kent.

RORISON, I. H. (1960), 'Some experimental aspects of the calcicole–calcifuge problem', *J. Ecol.*, **48**, 585–599.

RORISON, I. H. (1968), 'Ecological inferences from laboratory experiments on mineral nutrition', *Symp. Br. ecol. Soc.*, no. 9, 155–175.

ROSE, F. (1965), 'Comparaison phytogéographique entre les pelouses crayeuses du Meso-Xerobromion des vallées de la Basse-Seine, de la Somme, de l'Authie, de la Canche, de la Cuesta Boulonnais du Pas-de-Calais et du Sud Est de l'Angleterre', *Revue Socs sav. hte Normandie*, **37**, 105–109.

ROSE, F. (1972), 'Floristic connections between S.E. England and North France', in *Taxonomy, phytogeography and evolution*, ed. Valentine, D. H., 363–379. Proceedings of Symposium held by the Linnean Society of London, the Botanical Society of the British Isles and the International Organization of Plant Biosystematics, Manchester 1971. Academic Press, London.

ROUGHTON, R. D. (1972), 'Shrub age structure on a mule deer winter range in Colorado', *Ecology*, **53**, 615–625.

ROWLEY, T. (1972), *The Shropshire landscape*, Hodder and Stoughton, London.

SALISBURY, E. J. (1918), 'The ecology of scrub in Hertfordshire', *Trans. Herts. nat. Hist. Soc. Fld Club*, **17**, 53–64.

SALISBURY, E. J. (1920), 'The significance of the calcicolous habit', *J. Ecol.*, **8**, 202–215.

SALT, G., HOLLICK, F. S. J., RAW, F. and BRIAN, M. V. (1948), 'The arthropod population of pasture soil', *J. Anim. Ecol.*, **17**, 139–150.

SANDERSON, J. L. (1958), *Autecology of Corylus avellana L. in the neighbourhood of Sheffield with special reference to its regeneration*, Ph.D. thesis, University of Sheffield.

SCHOFIELD, R. K. and PENMAN, H. L. (1949), *The principles governing transpiration by vegetation*, Inst. Civ. Engrs Proc. Conf. Biol. Civ. Engng 1948, 75–84.

SCHRODER, D. and ZWÖLFER, H. (1969), 'Studies on insects associated with gorse, *Ulex europaeus* L., in *Proceedings of the first international symposium on biological control of weeds*, 55–58.

SCORER, A. G. (1913), *The entomologist's log-book*, Routledge, London.

SEARS, P. D. (1956), 'The effect of the grazing animal on pasture', *Int. Grassld Congr.*, 7th, 92–101.

SEARS, P. D. and NEWBOLD, R. P. (1942), 'The effect of sheep droppings on yield, botanical composition and chemical composition of pastures', *N.Z. Jl Sci. Technol.*, **24**, 36–61.

SHARP, W. E. (1915), *Common beetles of our countryside*, Partridge, London.

SHEAIL, J. (1971a), 'The formation and maintenance of water-meadows in Hampshire, England', *Biol. Conserv.*, **3**, 101–106.

SHEAIL, J. (1971b), *Rabbits and their history*, David and Charles, Newton Abbot.

SHEAIL, J. (1973), 'Changes in the use and management of farmland in England and Wales, 1915–19', *Trans. Inst. Br. Geogr.*, no. 60, 17–32.

SHEALS, J. G. (1957), 'The Collembola and Acarina of uncultivated soil', *J. Anim. Ecol.*, **26**, 125–134.

SHIMWELL, D. W. (1968), *The vegetation of the Derbyshire dales*, A report to the Nature Conservancy.

SHIMWELL, D. W. (1971a), 'Festuco–Brometea Br.-Bl. and R. Tx. 1943 in the British Isles: The phytogeography and phytosociology of limestone grasslands. Part I (a) General Introduction; (b) Xerobromion in England', *Vegetatio*, **23**, 1–28.

SHIMWELL, D. W. (1971b), 'Festuco–Brometea Br.-Bl. and R. Tx. 1943 in the British Isles: The phytogeography and phytosociology of limestone grasslands. Part II. Eu-Mesobromion in the British Isles', *Vegetatio*, **23**, 29–60.

SHIMWELL, D. W. (1971c), *The description and classification of vegetation*, Sidgwick and Jackson, London.

SHIMWELL, D. W. (1973), 'An introduction to the geography and ecology of chalk grassland', in *Chalk Grassland: Studies on its Conservation and Management in South East England*, eds. Jermy, A. C. and Stott, P. A., 1–5. Spec. Publs. Kent Trust for Nature Conservation.

SHOTTON, F. W. (1965), 'Movements of insect populations in the British Pleistocene', *Spec. Pap. geol. Soc. Am.*, **84**, 17–33.

SHOTTON, F. W. (1970), 'Quaternary entomology', *Proc. Coventry Distr. nat. Hist. scient. Soc.*, **4**(4), 101–109.

SHOTTON, F. W. and OSBORNE, P. J. (1965), 'The fauna of the Hoxnian interglacial deposits at Nechells, Birmingham', *Phil. Trans. R. Soc.*, B, **248**, 353–378.

SIDE, K. C. (1955), 'A study of the insects living on the wayfaring tree', *Bull. amat. Ent. Soc.*, **14**, 3–50.

SIMKHOVITCH, V. G. (1913), 'Hay and history', *Political Science Quarterly*, **28**, 385–404.

SIMMONS, I. G. (1964), 'Pollen diagram from Dartmoor', *New Phytol.*, **63**, 165–180.

SKILBECK, D. (ed.) (1956), 'Sheep farming in Romney Marsh in the 18th century', *Occ. Publs Wye Coll.*, no. 7.

SMITH, C. C. (1971), 'The coevolution of pine squirrels (*Tamiasciurus*) and conifers', *Ecol. Monogr.*, **40**, 349–371.

SMITH, C. J., ELSTON, J. and BUNTING, A. H. (1971), 'The effects of cutting and fertiliser treatments on the yield and botanical composition of chalk turf', *J. Br. Grassld Soc.*, **26**, 213–219.

SNAYDON, R. W. and BRADSHAW, A. D. (1969), 'Differences between material populations of *Trifolium repens* L. in response to mineral nutrients. II. Calcium, magnesium and potassium', *J. appl. Ecol.*, **6**, 185–202.

SNEDECOR, G. W. and COCHRAN, W. G. (1967), *Statistical methods*, Iowa State University Press, Iowa, USA.

SOCIETY FOR THE PROMOTION OF NATURE RESERVES CONSERVATION LIAISON COMMITTEE (1969), Biological sites recording scheme, *Tech. Publ. Soc. Promot. Nat. Reserves*, no. 1.

SOLOMON, M. E. (1971), 'Elements in the development of population dynamics', in *Dynamics of populations*, eds den Boer, P. J. and Gradwell, G. R., 29–40, Centre for Agricultural Publishing and Documentation, Wageningen.

SOUTHERN, H. N. (1964), *The handbook of British mammals*, Blackwell Scientific Publications, Oxford.

SOUTHWOOD, T. R. E. (1957), 'The zoogeography of the British Hemiptera–Heteroptera', *Proc. Trans. S. Lond. ent. nat. Hist. Soc.*, 1956, 111–136.

SOUTHWOOD, T. R. E. (1960), 'The flight activity of Heteroptera', *Trans. R. ent. Soc. Lond.*, **112**, 173–220.

SOUTHWOOD, T. R. E. (1961), 'The number of species of insect associated with various trees', *J. Anim. Ecol.*, **30**, 1–8.

SOUTHWOOD, T. R. E. (1966), *Ecological methods with particular reference to the study of insect populations*, Methuen, London.

SOUTHWOOD, T. R. E. (1972), 'The insect/plant relationship – an evolutionary perspective', *Symp. R. entomol. Soc. Lond.*, no. 6, 157–178.

SOUTHWOOD, T. R. E. and JEPSON, W. F. (1962), 'The productivity of grasslands in England for *Oscinella frit* (L.) (Chloropidae) and other stem-boring Diptera', *Bull. ent. Res.*, **53**, 395–407.

SOUTHWOOD, T. R. E. and LESTON, D. (1959), *Land and water bugs of the British Isles*, Warne, London.

SOUTHWOOD, T. R. E. and VAN EMDEN, H. F. (1967), 'A comparison of the fauna of cut and uncut grasslands', *Z. angew. Ent.*, **60**, 188–198.

SPARKS, B. W. (1964), 'Non-marine mollusca and quaternary ecology', *J. Anim. Ecol.*, **33** (Suppl.), 87–98.

SPENCER, K. A. (1972), 'Diptera, Agromyzidae', *Handbk Ident. Br. Insects*, **10** (5g), 1–136.

SPOONER, G. M. (1963), 'On causes of the decline of *Maculinea arion* L. (Lep. Lycaenidea) in Britain', *Entomologist*, **96**, 199–210.

STAMP, L. D. (1960), *Applied geography*, Penguin, London.

STAMP, L. D. (1962), *The land of Britain*, Longman, London.

STAPLEDON, R. G. (1937), *The hill lands of Britain*, Faber and Faber, London.

STAPLEDON, R. G. (1939), *The plough-up policy and ley farming*, Faber, London.

STAPLES, M. J. C. (1970), A history of box in the British Isles, *Boxwood Bulletin*, **10**, 18–23, 34–37 and 55–60.

STEVENS, JOSEPH (1874), *The farm labourer, at home, and in the field*.

STONE, THOMAS (1800), *A review of the corrected agricultural survey of Lincolnshire*.

STOTT, P. A. (1970), 'The study of chalk grassland in Northern France. An historical review', *Biol. J. Linnean Soc. Lond.*, **2**, 173–207.

STOUTJESDIJK, P. H. (1961), 'Micrometeorological measurements in vegetations of various structure I–III', *Proc. K. ned. Akad. Wet.*, 64C, **2**, 171–207.

STUBBS, A. E. (1972), 'Wild life conservation and dead wood', *J. Devon Trust Nature Conserv.*, Suppl.

STUTTARD, P. and WILLIAMSON, K. (1971), 'Habitat requirements of the nightingale', *Bird Study*, **18**, 1, 9–14.

SUTTON, S. L. (1968), 'The population dynamics of *Trichoniscus pusillus* and *Philoscia muscorum* (Crustacea Oniscoidea) in limestone grassland', *J. Anim. Ecol.*, **37**, 425–444.

SUTTON, S. L. (1972), *Woodlice*, Ginn and Co. Ltd., London.

TAMM, C. O. (1948), 'Observations on reproduction and survival of some perennial herbs', *Bot. Notiser*, 1948, 305–321.

TAMM, C. O. (1972), 'Survival and flowering of some perennial herbs', *Oikos*, **23**, 23–28.

TANSLEY, A. G. (1917), 'Competition between *Galium saxatile* and *G. sylvestre*', *J. Ecol.*, **5**, 173–179.

TANSLEY, A. G. (1922), 'Studies of the vegetation of the English chalk. II Early stages of redevelopment of woody vegetation of chalk grassland', *J. Ecol.*, **10**, 168–177.

TANSLEY, A. G. (1939), *The British Isles and their vegetation*, Cambridge University Press, Cambridge.

TANSLEY, A. G. (1945), *Our heritage of wild nature*, Cambridge University Press, Cambridge.

TANSLEY, A. G. (1946), *Introduction to plant ecology*, Allen and Unwin, London.

TANSLEY, A. G. (1953), *The British Islands and their vegetation*, Vols I and II, Cambridge University Press, Cambridge.

TANSLEY, A. G. and ADAMSON, R. S. (1925), 'Studies of the vegetation of the English chalk. III. The chalk grasslands of the Hampshire–Sussex border', *J. Ecol.*, **13**, 177–223.

TANSLEY, A. G. and ADAMSON, R. S. (1926), 'Studies of the vegetation of the English chalk. IV. A preliminary survey of the chalk grasslands of the Sussex Downs', *J. Ecol.*, **14**, 1–32.

TAYLOR, C. C. (1966), 'Strip lynchets', *Antiquity*, **40**, 277–284.

TAYLOR, D. (1971), 'London's milk supply, 1850–1900', *Agric. Hist.*, **45**, 33–38.

TAYLOR, K. and MARKS, T. C. (1971), 'The influence of burning and grazing on the growth and development of *Rubus chamaemorus* L. in *Calluna–Eriophorum* bog', *Symp. Br. ecol. Soc.*, no. 11, 153–166.

TAYLOR, L. R. (1968), 'The Rothamsted insect survey', *Natur. Sci. Sch.*, **6** (1), 2–9.

TESTER, J. R. and MARSHALL, W. H. (1961), 'A study of certain plant and animal interrelations on a native prairie in northwestern Minnesota', *Occ. Pap. Univ. Minn. Mus. nat. Hist.*, **8**, 51 pp.

THIRSK, J. (1965), *Fenland farming in the sixteenth century*, Department of English Local History, Leicester University Press, Occasional Paper No. 3.

THOMAS, A. S. (1959), 'Sheep paths', *J. Br. Grassld Soc.*, **14**, 157–164.

THOMAS, A. S. (1960), 'The tramping animal', *J. Br. Grassld Soc.*, **15**, 89–93.

THOMPSON, F. M. L. (1968), 'The second agricultural revolution, 1815–1880', *Econ. Hist. Rev.*, **21**, 62–77.

THORPE, H. (1965), 'The lord and the landscape', *Trans. Proc. Birmingham archaeol. Soc.*, **80**, 38–77.

TRACZYK, T. (1968), 'Studies on the primary production in meadow community', *Ekol. pol.*, Series A, **16**, 59–100.

TRIBE, D. E. (1950), 'The composition of a sheep's natural diet', *J. Br. Grassld Soc.*, **5**, 81–91.

TRIST, P. J. O. (1970), 'The changing pattern of agriculture', in *The flora of a changing Britain*, ed. Perring, F. H. Report of the Botanical Society of the British Isles, no. 11. Classey, Hampton (Middlesex).

TROELS-SMITH, J. (1960), 'Ivy, mistletoe and elm: climatic indicators – fodder plants', *Danm. geol. Unders.*, **4**, 1–32.

TUBBS, C. R. and JONES, E. L. (1964), 'The distribution of gorse (*Ulex europaeus* L.) in the New Forest in relation to former land use', *Proc. Hampshire Fld Club*, **23**, 1–10.

TURNER, J. (1962), 'The *Tilia* decline: an anthropogenic interpretation', *New Phytol.*, **61**, 328–341.

TURNER, J. (1965), 'A contribution to the history of forest clearance', *Proc. R. Soc.*, B, **161**, 343–354.

TURNOCK, W. J. (1972), 'Geographical and historical variability in population patterns and life systems of the larch sawfly', *Can. Ent.*, **104**, 1883–1900.

USHER, M. B. (1973), *Biological management and conservation*, Chapman and Hall, London.

VAN BATH, B. H. S. (1963), *The agrarian history of western Europe, AD 500–1850*, Arnold, London.

VAN DER MAAREL, E. and WESTHOFF, V. (1964), 'The vegetation of the dunes near Oostvoorne (The Netherlands) with a vegetation map', *Wentia*, **12**, 1–61.

VANDERMEER, J. H. (1972), 'Niche theory', *A. Rev. Ecol. and Syst.*, **3**, 107–132.

VAN EMDEN, H. F. and WAY, M. J. (1972), 'Host plants in the population dynamics of insects', *Symp. R. entomol. Soc. Lond.*, no. 6, 181–199.

VARLEY, G. C. (1947), 'The natural control of population balance in the knapweed gall-fly (*Urophora jaceana*)', *J. Anim. Ecol.*, **16**, 139–187.

VEALE, E. M. (1957), 'The rabbit in England', *Agric. Hist. Rev.*, **5**, 85–90.

VEDEL, H. (1961), 'Natural regeneration in juniper', *Proc. bot. Soc. Br. Isl.*, **4**, 146–148.

VENABLES, L. S. V. (1939), 'Bird distribution on the South Downs and a comparison with that of Surrey Greensand heaths', *J. Anim. Ecol.*, **8**, 227–237.

VOISIN, A. (1960), *Better grassland sward*, Crosby Lockwood, London.

VOISIN, A. (1961), *Grass productivity*, Crosby Lockwood, London.

WALKER, D. and WEST, R. G. (eds) (1970), *Studies in the vegetational history of the British Isles*, Cambridge University Press, Cambridge.

WALKER, S. S. (1920), 'The effect of aeration and other factors on the lime requirement of a mulch soil', *Soil Sci.*, **9**, 77–81.

WALLWORK, J. A. (1967), 'Acarina', in *Soil biology*, eds. Burges, A. and Raw, F., 363–395, Academic Press, London.

WALOFF, N. (1968), 'Studies on the insect fauna on scotch broom, *Sarothamnus scoparius* (L.) Wimmer', *Adv. Ecol. Res.*, **5**, 87–208.

WALOFF, N. and SOLOMON, M. G. (1973), 'Leafhoppers (Auchenorhyncha Homoptera) of acid grassland', *J. appl. Ecol.*, **10**, 189–212.

WARD, L. K. (1966), *The biology of Thysanoptera*. Ph.D. thesis, University of London.

WARD, L. K. (1973), 'The conservation of juniper. I. Present status of juniper in southern England', *J. appl. Ecol.*, **10**, 165–188.

WATERHOUSE, F. L. (1955), 'Microclimatological profiles in grass cover in relation to biological problems', *Q. Jl R. met. Soc.*, **81**, 63–71.

WATERSTON, A. R. (1964), 'On *Zygimus nigriceps* (Fallen, 1829) (Hem., Miridae) and its host-plant', *Entomologist*, **97**, 248–249.

WATKIN, B. R. (1954), 'The animal factor and levels of nitrogen', *J. Br. Grassld Soc.*, **9**, 35–46.

WATKIN, B. R. (1957), 'The effect of dung and urine and its interactions with applied nitrogen, phosphorus and potassium on the chemical composition of pasture', *J. Br. Grassld Soc.*, **12**, 264–277.

WATSON, A. (1967), 'Social status and population regulation in the red grouse (*Lagopus lagopus scoticus*)', *Proc. R. Soc. Population Study Group*, **2**, 22–30.

WATSON, H. C. (1847), *Cybele Britannica*, London.

WATT, A. S. (1926), 'Yew communities of the South Downs', *J. Ecol.*, **14**, 282–316.

WATT, A. S. (1934), 'The vegetation of the Chiltern Hills with special reference to the beech-woods and their seral relationships', *J. Ecol.*, **22**, 230–270 and 445–507.

WATT, A. S. (1938), 'Studies on the ecology of Breckland. III. The origin and development of the Festuco–Agrostidetum on eroded sand', *J. Ecol.*, **26**, 1–37.

WATT, A. S. (1947), 'Pattern and process in the plant community, *J. Ecol.*, **35**, 1–22.

WATT, A. S. (1971), 'Factors controlling the floristic composition of some plant communities in Breckland', *Symp. Br. Ecol. Soc.*, no. 11, 137–152.

WATT, K. E. F. (1966), *Systems analysis in ecology*, Academic Press, London.

WATTS, C. H. S. (1968), 'The food eaten by wood mice (*Apodemus sylvaticus*) and bank voles (*Clethrionomys glareolus*) in Wytham Woods, Berkshire', *J. Anim. Ecol.*, **37**, 25–41.

WAY, J. M. (ed.) (1969), *Road verges: their function and management*, Monks Wood Experimental Station. The Nature Conservancy.

WEBB, D. A. (1954), 'Is the classification of vegetation either possible or desirable?' *Bot. Tidsskr.*, **51**, 362–370.

WELCH, D. and RAWES, M. (1966), 'The intensity of sheep grazing on high-level blanket bog in upper Teesdale', *Ir. J. agric. Res.*, **5**, 185–196.

WELCH, R. C. (1964), 'A simple method of collecting insects from rabbit burrows', *Entomologist's mon. Mag.*, **100**, 99–100.

WELCH, R. C. (1965), *The biology of the genus* Aleochara *Grav. (Coleoptera (Staphylinidae))*, Ph.D. thesis, University of London.

WELCH, R. C. (1972), 'The biology of *Hermaeophaga mercurialis* F. (Coleoptera, Chrysomelidae)', *Entomologist's Gaz.*, **23**, 153–166.

WELLHOUSE, W. H. (1922), 'The insect fauna of the genus *Crataegus*', *Mem. Cornell Univ. agric. Exp. Stn*, **56**, 1041–1136.

WELLS, T. C. E. (1965), 'Chalk grassland nature reserves and their management problems', *Handbk a. Rep. Soc. Promot. Nat. Reserves*, 1965, 62–70.

WELLS, T. C. E. (1967), 'Changes in a population of *Spiranthes spiralis* (L.) Chevall at Knocking Hoe National Nature Reserve, Bedfordshire, 1962–65', *J. Ecol.*, **55**, 83–99.

WELLS, T. C. E. (1971), 'A comparison of the effects of sheep grazing and mechanical cutting on the structure and botanical composition of chalk grassland', *Symp. Br. ecol. Soc.*, no. 11, 497–515.

WELLS, T. C. E. and BARLING, D. M. (1971), 'Biological flora of the British Isles: *Pulsatilla vulgaris* Mill', *J. Ecol.*, **59**, 275–292.

WEST, O. (1965), *Fire in vegetation and its use in pasture management, with special reference to tropical and subtropical Africa*, Commonwealth Agricultural Bureaux, Farnham Royal.

WESTHOFF, V. (1952), 'Gezelschappen met Loutigue gewassen in de duinen en langs de Binnemduinrand', *Jaarb. ned. dendrol. Vereen. 1950–1*, **18**, 9–49.

WESTHOFF, V. and DEN HELD, A. H. (1969), *Platen–Gemeeenschappen in Nederland*, N.V.W.J. Thieme and CIE, Zutphen.

WHITMAN, W. C. (1969), 'Microclimate and its importance in grassland ecosystems', in *The grassland ecosystem: a preliminary synthesis*, eds Dix, R. L. and Beideleman, R. G., Colorado State University, Colorado.

WHITTAKER, E. (1960), *Ecological effects of moor burning*, Ph.D. thesis, University of Aberdeen.

WHITTAKER, J. B. (1969), 'Quantitative and habitat studies of the frog-hoppers and leaf-hoppers (Homoptera, Auchenorhyncha) of Wytham Woods, Berkshire', *Entomologist's mon. Mag.*, **105**, 27–37.

WHITTAKER, R. H. (1970), *Communities and ecosystems*, Macmillan, London and Toronto.

WILCOX, J. C. and SPILSBURY, R. H. (1941), 'Soil moisture studies', *Scient. Agric.*, **21**, 459–472.

WILLIAMS, C. B. (1964), *Patterns in the balance of nature and related problems in quantitative biology*, Academic Press, London.

WILLIAMS, J. T. (1968), 'The nitrogen relations and ecological investigations on wet fertilised meadows', *Veröff. geobot. Inst. Zurich*, **41**, 70–193.

WILLIAMS, J. T. and VARLEY, Y. W. (1967), 'Phytosociological studies of some British grasslands', *Vegetatio*, **15**, 169–189.

WILLIAMS, M. (1972), 'The enclosure of waste land in Somerset, 1700–1900', *Trans. Inst. Br. Geogr.*, **57**, 99–124.

WILLIAMS, O. B., WELLS, T. C. E. and WELLS, D. A. (1974), 'Grazing management of Woodwalton Fen: seasonal changes in the diet of cattle and rabbits', *J. appl. Ecol.* **11** (2).

WILLIAMSON, K. (1967), 'Some aspects of the scientific interest and management of scrub on nature reserves, in *The biotic effects of public pressures on the environment*, ed. Duffey, E. 94–100, Monks Wood Experimental Station Symposium No. 3, The Nature Conservancy.

WILLIAMSON, K. and WILLIAMSON, R. (1973), 'The bird community of yew woodland at Kingley Vale, Sussex', *Br. Birds*, **66**, 12–23.

YATES, F. (1937), 'The design and analysis of factorial experiments', *Tech. Commun. Commonw. Bur. Soil Sci.*, **35**, 1–95.

YOUNG, ARTHUR (1813), *General view of the agriculture of Oxfordshire*, Board of Agriculture, London.

Index

Index

This index should be used in conjunction with the Contents list (pp. v–ix). Plants and animals (except where the common name is more generally used, e.g. in birds) are entered under the Latin names only. In most cases where several species are mentioned in the text, the genus only is given. Figures in bold type refer to plates.